Phonetica

Constancy and Variation in Speech

Editors
Calbert R. Graham, Cambridge
Brechtje Post, Cambridge

37 figures and 17 tables, 2019

KARGER Basel · Freiburg · Hartford · Oxford · Bangkok · Dubai · Kuala Lumpur · Melbourne · Mexico City · Moscow · New Delhi · Paris · Shanghai · Tokyo

S. Karger
Medical and Scientific Publishers
Basel · Freiburg · Hartford · Oxford ·
Bangkok · Dubai · Kuala Lumpur ·
Melbourne · Mexico City ·
Moscow · New Delhi · Paris ·
Shanghai · Tokyo

Printed on acid-free and non-aging paper
(ISO 9706)
ISBN 978–3–318–06474–2
e-ISBN 978–3–318–06475–9

KARGER

E-Mail karger@karger.com
www.karger.com/pho

Vol. 76, No. 2–3, 2019

Phonetica

Contents

KARGER

E-Mail karger@karger.com
www.karger.com/pho

Introduction

Phonetica 2019;76:87–99
DOI: 10.1159/000497439

Received: October 29, 2018
Accepted: January 20, 2019
Published online: May 21, 2019

Constancy and Variation in Speech: Phonetic Realisation and Abstraction

Calbert Graham Brechtje Post

Phonetics Laboratory, Theoretical and Applied Linguistics, University of Cambridge, Cambridge, UK

The mechanisms involved in the production of a speech sound vary from one speaker or iteration to the next, resulting in considerable variation in the phonetic realisation of speech sounds. The exact settings of the articulators in the production of a speech sound will vary across speakers, but, even within the same speaker, no two repetitions of a sound are quite the same. Furthermore, the phonetic realisation of a common form may vary according to an individual's dialect, speaking style, or even in the way sounds coarticulate or overlap with each other in connected speech. Yet, despite the intrinsically gradient nature of phonetic variation, listeners are usually successful in working out the intended message of the speaker by extracting meaningful categories from the speech signal. It goes without saying, therefore, that interpreting the intention of the speaker along with the speech form he/she produces in a given context is central to the process of mapping between speech segments and abstract representations. Put another way, the communicative use of language necessarily involves extracting "constancy," in the sense of category membership, from "variation" in any given context.

However, the listener's task of abstracting constants or "constancy" from multiple sources of variation (both volitional and intrinsic) created by the population of speakers encountered is by no means a trivial one. We do not know nearly enough about how this process works or about the operations of its underlying mechanisms. Moreover, the process of interpreting variation in order to arrive at the speaker's intended meaning does not always occur seamlessly, since listeners can fail at correctly interpreting the signal. Therefore, our understanding of the relationship between constancy and variation hinges on unravelling the complex interplay of the many factors (social, cognitive, linguistic, biological, environmental, and so on) that determine the mapping between them.

This complexity of mapping constitutes one of the classical challenges for the speech sciences, and it forms the topic of this special issue. Going back to the

KARGER

E-Mail karger@karger.com
www.karger.com/pho

Dr. Calbert Graham
Phonetics Laboratory
University of Cambridge
Cambridge CB3 9DA (UK)
E-Mail crg29@cam.ac.uk

seminal work by Ladefoged and Broadbent (1957), early work sought to identify the units on which the process of encoding or decoding phonetic information operates. It also sought to model the mechanisms that supported that encoding or decoding (e.g., Klatt, 1979; Liberman & Mattingly, 1985), yielding a plethora of theoretical proposals. Among the many proposed concepts is (relational) *invariance*, which assumes that distinctive features are the primary perceptual unit in the sound system of a language (e.g., Stevens, 2002, based on ideas inherent in earlier work such as Jakobson et al., 1963; Liberman et al., 1967; Pisoni, 1978; Stevens & House, 1972). Another is *compensation for coarticulation,* which addresses the strategies speakers use in natural speech when they are unable to reach optimal places of articulation to achieve an auditory contrast (e.g., Mann, 1980; Mann & Repp, 1981). Further, in the *cohort model* speech input is progressively mapped onto competing lexical forms that are evaluated against one another (Marslen-Wilson, 1987). Finally, another relevant model is built upon *exemplar-based theories* in which linguistic knowledge is built by the memory of an individual speaker-hearer's complete linguistic experiences, shifting attention away from abstractions (inspired by exemplar models of categorisation: Goldinger, 1996; Hintzman, 1986; Kruschke, 1992; Nosofsky, 1986).

In parallel, a rapidly growing literature started to examine the sources of variation, initially asking questions about issues such as (1) the auditory and articulatory constraints on speech perception and production (e.g., Browman & Goldstein, 1992; Lindblom, 1990; Miller, 1989; Stevens, 1989), (2) the effects of coarticulatory and other connected speech processes on units of speech (e.g., Mann, 1980; Nolan, 1983), (3) the origins of sound change (e.g., Blevins, 2004; Ohala, 1974, 1981), and (4) the nature of dialectal diversity due to social factors and regional variation (e.g., Hay et al., 2006; Niedzielski, 1999; Strand, 1999). For instance, in their work on sociophonetic variation, Foulkes and Docherty (2006) identify speaker gender, age, ethnicity, social class, group affiliations, geographical origin, and speaking style as sources – all of which had long been identified but without the supporting experimental evidence yielded by more recent studies.

This exploration of the multiple sources of variation highlights that the performance of individual speakers/listeners must be factored into accounts of speech production, perception, acquisition, and diachronic change (Docherty & Foulkes, 2014). It will also lead to an increased awareness that phonetic realisation and abstraction can only be fully understood if the full gamut of non-linguistic factors is taken into consideration, calling for a more interdisciplinary approach. Hawkins (2003, p. 373), for instance, argued that significant progress can be made "by changing our focus of inquiry from abstraction of formal units of linguistic analysis to a detailed analysis of global aspects of the communicative situation, of which speech is just one part" (see also Hawkins, 2010; Hawkins & Smith, 2001; Local, 2003; and Best, 2015, who suggests that successful word recognition requires perceivers to exploit complimentary principles of phonological distinctiveness and phonological constancy – see section "Perception of accent" below for a brief discussion of this idea in the context of lexical processing).

These insights have opened up new avenues for research on constancy and variation in speech involving a wide range of disciplines beyond phonetics. For instance, in arguing that theoretical accounts of the effect of adverse listening conditions on speech recognition must include information about the environment as opposed to trying to normalise it, Mattys et al. (2012) highlight the wider usefulness of adverse listening research for work on fundamental psycholinguistic topics such as the structure of lexical representations, segmentation, and feedback in perception.

This special issue seeks to take stock of these latest theoretical developments and empirical advances by moving the focus away from questions about the nature of the constancy and the variation that is at play (see e.g. the special issues edited by Hawkins & Nguyen, 2003, in *Journal of Phonetics* and by Vigário et al., 2015, in *Phonetica*) to questions about the effects of increased variability on the relationship between constancy and variation. The contributions to this issue show how increased variability due to the interplay of multiple sources can facilitate perception and affect sound change. We turn now to discuss each of the papers in this special issue, grouped under two themes: the interplay of multiple sources of variability, and variability at the tipping point, which we define as the point at which surface variation is no longer interpreted as phonetic variation, but as an instance of a different "constant" (diachronically or in online speech perception).

The Interplay of Multiple Sources of Variability

Sound Change

The perceptual processing and the effects of the variations that arise from language change and language contact more broadly contribute significantly to sound change. This makes it imperative for us to consider sound change as the backdrop to examining the nature and source of these effects. Yet, despite the sizeable body of research on phonetically-based sound change, there are still several outstanding issues that need to be elucidated if we are to have a satisfactory understanding of the roles of the various mechanisms of sound change. Some key questions include: What is the extent of variation in speaker productions due to non-phonetic factors? To what degree is production affected by perception? Are there variables that can be considered markers of sound change in action? These questions are addressed directly in this special issue by Karolina Broś and Katarzyna Lipowska in their paper "Grand Canarian Spanish Non-Continuant Voicing: Gradiency, Sex Differences and Perception." They take a Labovian (Labov, 1972) or variationist (Guy, 1994) point of view, in which the relationship between synchronic variation and diachronic sound change depends on the complex interplay between various sources of change, both linguistic and extralinguistic. On the basis of a detailed analysis of post-vocalic voicing in Gran Canarian Spanish, the authors conclude that, far from being a unitary internal phonetic process, variation that underpins sound change very much depends on social factors such as gender, education, and individual communicative stances. The key finding of this study serves as a unifying statement that brings together two different strands of sound change research: on the one hand, there are those who are concerned with universal patterns in sound change that arise through the physical (articulatory and perceptual) processes related to the speech pro-

Phonetica 2019;76:87–99
DOI: 10.1159/000497439

duction and perception apparatus. This strand has been largely developed around the work of John Ohala mentioned above. Then, on the other hand, there is the work that focusses on social and environmental factors in sound change, such as the line of research on gender effects in sound change that has established, for example, that women's speech tends to lead to sound change (e.g., Eckert & McConnell-Ginet, 2003; Labov, 2001), while older men tend to be most conservative with respect to initiating or adopting speech changes (e.g., Orton, 1962). Broś and Lipowska's work specifically emphasises that a full understanding of the sources of sound change crucially depends on a better understanding of not only the individual effects of the many internal as well as external factors that play a role, but also of the way in which they interact over time.

Intelligibility of Second Language Speech

Recent research strongly suggests that a crucial factor in the ability to abstract and categorise speech when it is greatly variable – regardless of the source of that variability – is the amount and type of variability to which the listener is exposed. For instance, Logan and colleagues' research on "high variability phonetic training" (Logan et al., 1991) with non-native speakers has provided evidence that allophonic variation through the use of multiple speakers and varied segmental contexts facilitates categorical abstraction and leads to improved perception and production. The findings demonstrate that exposure to variability is a critical aspect of developing perceptual proficiency in a second language.

In their contribution to this special issue, Melissa Baese-Berk and Tuuli Morrill ("Perceptual Consequences of Variability in Native and Non-Native Speech") ask whether variability also affects intelligibility in a second language. Research on speech perception in second language learners has generally shown that listeners find utterances by native speakers of a language more intelligible than those spoken by second language learners (e.g., Bent & Bradlow, 2003; Munro, 1998; Munro & Derwing, 1999; Smith et al., 2003; van Wijngaarden, 2001). Bent and Bradlow (2003) also report evidence for a "matched interlanguage speech intelligibility benefit" wherein second language listeners rated other (high proficiency) second language English speakers from the same language background as themselves as being equally intelligible as native speakers. This finding confirms that an adequate understanding of speech intelligibility must take both talker- and listener-related factors into account. However, Baese-Berk et al. (2013) showed that listeners' ability to extract phonological constancy from phonetic variability is not necessarily a language-specific capacity, as they generalise their learning to novel speakers from language backgrounds both included and not included in the perceptual training they receive. Essentially, these studies highlight the complex interplay of the factors involved.

In this special issue, Baese-Berk and Morrill take up this issue by addressing the question of how a specific type of variability (speaking rate) impacts intelligibility and fluency (as judged by listener ratings). They find that it is not only the case that listeners with more experience of more variable speech show better perception of non-native speech (Laturnus, 2018), but their perceptual ability also depends on the type of variability to which they have been exposed. Note, however, that other sources of variability can also affect an individual's performance, such as their proficiency in their target L2 and the similarity be-

DOI: 10.1159/000497439

tween their target L2 and their native language (Graham & Post, 2018). The relationship between intelligibility and variability by native and non-native listeners/speakers is clearly a complex one. More research is needed that focusses on different phonetic measures of intelligibility as well as removing an important confound by teasing apart the notion of intelligibility from fluency. Whilst there is an appreciable amount of research on how second language listeners perceive the utterances of their target L2, there is not much research on how utterances produced by second language speakers are perceived by different populations of listeners.

The authors conclude that the results of their study underscore the need for the development or adaptation of current models to account for the way speech utterances produced by second language speakers are perceived by different categories of listeners. By refining models of speech perception to include different sources of variability and how they impact on perception, we will be better able to understand the factors that contribute to listeners' ability to comprehend speech utterances produced by lower proficiency second language learners and determine what compensatory strategies listeners use to decode the intended message. This line of work has important implications for both human as well as machine processing of second language speech utterances.

Perception of Accent

In their contribution to the special issue, Bronwen Evans, Amy Madigan, and Gisela Lourido investigate whether greater exposure to variable phonetic input for bilingual children in a multilingual community facilitates their perception of accent in "Effects of Language Background on the Development of Sociolinguistic Awareness." We know from studies with monolingual children that exposure to variable phonetic input plays a crucial role in acquiring their native language. Werker and Tees (1984) showed that infants can discriminate non-native speech contrasts without any previous exposure to the languages, and that this ability declines around the end of the first year of life. This decline has been posited to signal a shift from phonetic-based perception to the acquisition of phonological contrasts and appears to be related to the increase in exposure to the native language. That relationship is corroborated in later research by Nathan et al. (1998) who found that more experience with variation results in better ability to retrieve constant phonological and lexical information, which should be affected by the relative frequency of different types of variation (e.g., most people hear more gender-related variation than non-native dialectal variation). The question of how children develop this perceptual awareness has been addressed by Pierrehumbert (2003) who hypothesised that phonological categorisation is a bottom-up process based on the ongoing statistical analysis of density distributions in a phonetic space. This hypothesised process constitutes a statistical learning mechanism that relates to exemplar models. By Pierrehumbert's account, it is from this mechanism that abstract, phonological categories would emerge. However, Best et al. (2009) showed that whilst 15- and 19-month-old children are able to recognise familiar words spoken in their native dialect, only the 19-month-old children are able to recognise familiar words spoken in an unfamiliar non-native (Jamaican) dialect. Those researchers suggest that acquisition of phonological constancy emerges around that age. They further ar-

Phonetica 2019;76:87–99
DOI: 10.1159/000497439

gue that their results are incompatible with a statistical learning account of developmental change in early word recognition given that the older children are generalising readily to an accent they have not previously experienced, and instead support a perceptual-attunement account (as proposed in the Perceptual Assimilation Model; Best, 1995). This area of inquiry has implications for adults as well; Graham and Williams (2016), for example, showed that certain phonological abstraction processes continue to function during adulthood. In their study, adult second language learners (Japanese native listeners) were able to implicitly acquire knowledge of the phonological patterns (lexical word stress and phonotactic rules) of a novel second language through a brief auditory experience and without any conscious awareness of the rules they were acquiring.

The relationship between phonological constancy and phonetic variation has also been examined within the context of bilingual speech acquisition. Previous studies of bilingual children show that their greater exposure to phonetic variation and associated phonological categories in two languages can give them an advantage over monolingual children in the acquisition of certain aspects of speech (e.g., Schmidt & Post, 2015). In this special issue, Evans and colleagues use the varied language background of migrant bilingual children in London to establish whether early experience with variation affects the ability to categorise different accents: home versus unfamiliar regional accent, home versus unfamiliar foreign accent, and unfamiliar regional versus unfamiliar foreign accent. The authors find that bilinguals were better able to categorise talkers in all accent conditions than monolingual speakers. One explanation the authors offer for this finding is that, in addition to exposure to their own native dialect, bilinguals are aided by greater exposure than monolinguals to other sources of variation, such as the less predictable differences that occur in foreign-accented speech (Girard, Floccia and Goslin, 2008). More generally, the authors argue that knowledge of phonetic and social variation is likely emergent with the development of the ability to extract and use patterns of variation to categorise talkers and is highly dependent on linguistic experience. This underscores the importance of sociolinguistic awareness in the processing of phonetic variation and sheds light on how children acquire and use this knowledge to process phonetic variation.

Lexical Processing

Social factors have also been found to play a role in speech processing. It has been argued that the associated phonetic variation should not be dealt with in a pre-processing normalisation step that precedes categorisation, as is the case in classical models of speech perception, but instead by a fully integrated component of the speech perception system as a whole (e.g., Nusbaum and Magnuson, 1997). Nusbaum and Magnuson (1997) proposed that the system adapts its processing depending on certain variables – in their case largely talker and phonetic context variability – and operates in a way akin to hypothesis testing based on the prior context, listener expectations, and information in the signal. A recurring question in research on speech perception and word recognition is how useful small details of phonetic variation are and how they support listener attunement to phonological contrast in their native language. Best (2015) suggests that language learners are able to make use of organised sets of relation-

Foulkes, P., & Docherty, G. (2006). The social life of phonetics and phonology. *Journal of Phonetics, 34*(4), 409–438. doi:10.1016/j.wocn.2005.08.002

Girard F, Floccia C, Goslin J. (2008). Perception and awareness of accents in young children. *British Journal of Developmental Psychology, 26*, 409–433.

Goldinger, S. D. (1996). Words and voices: Episodic traces in spoken word identification and recognition memory. *Journal of Experimental Psychology. Learning, Memory, and Cognition, 22*(5), 1166–1183. doi:10.1037/0278-7393.22.5.1166

Goldinger, S. D. (1998). Echoes of echoes? An episodic theory of lexical access. *Psychological Review, 105*(2), 251–279. doi:10.1037/0033-295X.105.2.251

Graham, C., & Post, B. (2018). Second language acquisition of intonation: Peak alignment in American English. *Journal of Phonetics, 66*, 1–14. doi:10.1016/j.wocn.2017.08.002

Graham, C., & Williams, J. N. (2016). Implicit learning of Latin stress regularities. *Studies in Second Language Acquisition, 39*, 1–27.

Gureckis TM, Goldstone RL (2008): The effect of the internal structure of categories on perception; in Love BC, McRae K, Sloutsky VM (eds): *Proceedings of the 30th Annual Conference of the Cognitive Science Society*. Austin, Cognitive Science Society, pp 1876–1881.

Guy, G. R. (1994). The phonology of variation. In K. Beals (Ed.), *CLS 30: Vol. 2. Parasession on variation in linguistics theory* (pp. 133–149). Chicago, IL: Chicago Linguistic Society.

Hawkins, S. (2003). Roles and representations of systematic fine phonetic detail in speech understanding. *Journal of Phonetics, 31*(3-4), 373–405. doi:10.1016/j.wocn.2003.09.006

Hawkins, S. (2010). Phonetic variation as communicative system: Perception of the particular and the abstract. In C. Fougeron, B. Kuehnert, M. Imperio, & N. Vallee (Eds.), *Laboratory Phonology* (Vol. 10, pp. 479–510). Berlin: De Gruyter Mouton.

Hawkins, S., & Nguyen, N. (Eds.) (2003). Temporal integration in the perception of speech [Special Issue]. *Journal of Phonetics, 31*, 276–626.

Hawkins, S., & Smith, R. (2001). Polysp: A polysystemic, phonetically-rich approach to speech understanding. *Rivista di Linguistica, 13*, 99–188.

Hay, J., Warren, P., & Drager, K. (2006). Factors influencing speech perception in the context of a merger-in-progress. *Journal of Phonetics, 34*(4), 458–484. doi:10.1016/j.wocn.2005.10.001

Hintzman, D. L. (1986). Schema abstraction in a multiple-trace memory model. *Psychological Review, 93*(4), 411–428. doi:10.1037/0033-295X.93.4.411

Jakobson, R., Fant, G. M., & Halle, M. (1963). *Preliminaries to Speech Analysis: The Distinctive Features and their Correlates*. Cambridge: MIT Press.

Johnson, K. (1997). Speech perception without speaker normalization: An exemplar model. In K. Johnson & J. W. Mullennix (Eds.), *Talker Variability in Speech Processing* (pp. 145–165). San Diego: Academic Press.

Johnson, K. (2006). Resonance in an exemplar-based lexicon: The emergence of social identity and phonology. *Journal of Phonetics, 34*(4), 485–499. doi:10.1016/j.wocn.2005.08.004

Kapatsinski, V. (2018). *Changing Minds Changing Tools: From Learning Theory to Language Acquisition to Language Change*. Cambridge: MIT Press. doi:10.7551/mitpress/11400.001.0001

Kirby, J. (2010). *Cue selection and category restructuring in sound change* (PhD dissertation), University of Chicago.

Klatt, D. H. (1979). Speech perception: A model of acoustic-phonetic analysis and lexical access. *Journal of Phonetics, 7*, 279–312.

Kruschke, J. K. (1992). ALCOVE: An exemplar-based connectionist model of category learning. *Psychological Review, 99*(1), 22–44. doi:10.1037/0033-295X.99.1.22

Labov, W. (1972). *Sociolinguistic Patterns*. Philadelphia: University of Pennsylvania Press.

Labov, W. (2001). Principles of Linguistic Change.: Vol. 2. *Social factors*. Oxford: Blackwell.

Ladefoged, P., & Broadbent, D. E. (1957). Information conveyed by vowels. *The Journal of the Acoustical Society of America, 29*(1), 98–104. doi:10.1121/1.1908694

Laturnus, R. (2018). *Perceptual adaptation to non-native speech: The effects of bias, exposure, and input variation* (PhD dissertation), New York University.

Liberman, A. M., & Mattingly, I. G. (1985). The motor theory of speech perception revised. *Cognition, 21*(1), 1–36. doi:10.1016/0010-0277(85)90021-6

Liberman, A. M., Cooper, F. S., Shankweiler, D. P., & Studdert-Kennedy, M. (1967). Perception of the speech code. *Psychological Review, 74*(6), 431–461. doi:10.1037/h0020279

Lindblom, B. (1990). Explaining phonetic variation: A sketch of the H&H theory. In W. J. Hardcastle & A. Marchal (Eds.), *Speech Production and Speech Modelling* (pp. 403–439). Dordrecht: Springer Netherlands. doi:10.1007/978-94-009-2037-8_16

Local, J. (2003). Variable domains and variable relevance: Interpreting phonetic exponents. *Journal of Phonetics, 31*(3-4), 321–339. doi:10.1016/S0095-4470(03)00045-7

Logan, J. S., Lively, S. E., & Pisoni, D. B. (1991). Training Japanese listeners to identify English /r/ and /l/: A first report. *The Journal of the Acoustical Society of America, 89*(2), 874–886. doi:10.1121/1.1894649

Mann, V. A. (1980). Influence of preceding liquid on stop-consonant perception. *Perception & Psychophysics, 28*(5), 407–412. doi:10.3758/BF03204884

Mann, V. A., & Repp, B. H. (1981). Influence of preceding fricative on stop consonant perception. *The Journal of the Acoustical Society of America, 69*(2), 548–558. doi:10.1121/1.385483

Marslen-Wilson, W. D. (1987). Functional parallelism in spoken word-recognition. *Cognition, 25*(1-2), 71–102. doi:10.1016/0010-0277(87)90005-9

Mattys, S. L., Davis, M. H., Bradlow, A. R., & Scott, S. K. (2012). Speech recognition in adverse conditions: A review. *Language and Cognitive Processes, 27*(7-8), 953–978. doi:10.1080/01690965.2012.705006

Miller, J. D. (1989). Auditory-perceptual interpretation of the vowel. *The Journal of the Acoustical Society of America, 85*(5), 2114–2134. doi:10.1121/1.397862

Munro, M. J. (1998). The effects of noise on the intelligibility of foreign-accented speech. *Studies in Second Language Acquisition, 20*(02), 139–154. doi:10.1017/S0272263198002022

Munro, M. J., & Derwing, T. M. (1999). Foreign accent, comprehensibility, and intelligibility in the speech of second language learners. *Language Learning, 49*, 285–310. doi:10.1111/0023-8333.49.s1.8

Nathan, L., Wells, B., & Donlan, C. (1998). Children's comprehension of unfamiliar regional accents: A preliminary investigation. *Journal of Child Language, 25*(2), 343–365. doi:10.1017/S0305000998003444

Niedzielski, N. (1999). The effect of social information on the perception of sociolinguistic variables. *Journal of Language and Social Psychology, 18*(1), 62–85. doi:10.1177/0261927X99018001005

Nolan, F. (1983). *The Phonetic Bases of Speaker Recognition*. Cambridge: Cambridge University Press.

Norris, D., McQueen, J. M., & Cutler, A. (2003). Perceptual learning in speech. *Cognitive Psychology, 47*(2), 204–238. doi:10.1016/S0010-0285(03)00006-9

Nosofsky, R. M. (1986). Attention, similarity, and the identification-categorization relationship. *Journal of Experimental Psychology. General, 115*(1), 39–61. doi:10.1037/0096-3445.115.1.39

Nusbaum, H., & Magnuson, J. (1997). Talker normalization: Phonetic constancy as a cognitive process. In K. A. Johnson & J. W. Mullennix (Eds.), *Talker Variability and Speech Processing* (pp. 109–132). New York: Academic Press.

Ohala, J. J. (1974). Experimental historical phonology. In J. M. Anderson & C. Jones (Eds.), *Proceedings of the First International Conference on Historical Linguistics*. Edinburgh, 2–7 Sep 1973(pp. 353–389). Amsterdam, the Netherlands: North-Holland.

Ohala, J. J. (1981). The listener as a source of sound change. In C. S. Masek, R. A. Hendrick, & M. F. Miller (Eds.), *Papers from the Parasession on Language and Behavior* (pp. 178–203). Chicago: Chicago Linguistic Society.

Orton, H. (1962). *Survey of English Dialects: Introduction*. Leeds: EJ Arnold & Son.

Pierrehumbert, J. B. (2001). Exemplar dynamics: Word frequency, lenition and contrast. In J. L. Bybee & P. J. Hopper (Eds.), *Frequency and the Emergence of Linguistic Structure* (pp. 137–157). Amsterdam: John Benjamins. doi.org/10.1075/tsl.45.08pie

Pierrehumbert, J. B. (2003). Phonetic diversity, statistical learning, and acquisition of phonology. *Language and Speech, 46*(Pt 2-3), 115–154. doi:10.1177/00238309030460020501

Pisoni, D. B. (1978). Speech perception; in W. K. Estes (Ed.), *Handbook of learning and cognitive processes: Vol. 6. Linguistic functions in cognitive theory* (pp. 167–233). Oxford, UK: Erlbaum.

Samuel, A. G., & Kraljic, T. (2009). Perceptual learning for speech. *Attention, Perception & Psychophysics, 71*(6), 1207–1218. doi:10.3758/APP.71.6.1207

Scharenborg, O., Kakouros, S., Post, B., & Meunier, F. (2019). Cross-linguistic influences on sentence accent detection in background noise. *Language and Speech*, Epub ahead of print. doi:10.1177/0023830918819573

Schmidt, E., & Post, B. (2015). The development of prosodic features and their contribution to rhythm production in simultaneous bilinguals. *Language and Speech, 58*(Pt 1), 24–47. doi:10.1177/0023830914565809

Smith, B., Bradlow, A. R., & Bent, T. (2003). Production and perception of temporal contrasts in foreign-accented English. In M. J. Solé, D. Recasens, & J. Romero (Eds.), *Proceedings of the 15th International Congress on Phonetic Science* (pp. 519–522). Adelaide, Australia: Causal Productions.

Sóskuthy, M. (2015). Understanding change through stability: A computational study of sound change actuation. *Lingua, 163*, 40–60. doi:10.1016/j.lingua.2015.05.010
Stevens, K. N. (1989). On the quantal nature of speech. *Journal of Phonetics, 17*, 3–45.
Stevens, K. N. (2002). Toward a model for lexical access based on acoustic landmarks and distinctive features. *The Journal of the Acoustical Society of America, 111*(4), 1872–1891. doi:10.1121/1.1458026
Stevens, K. N., & House, A. S. (1972). Speech perception. In J. V. Tobias (Ed.), *Foundations of Modern Auditory Theory* (Vol. 11, pp. 1–62). New York: Academic Press.
Strand, E. A. (1999). Uncovering the role of gender stereotypes in speech perception. *Journal of Language and Social Psychology, 18*(1), 86–99. doi:10.1177/0261927X99018001006
Sumner, M., Kim, S. K., King, E., & McGowan, K. B. (2014). The socially weighted encoding of spoken words: A dual-route approach to speech perception. *Frontiers in Psychology, 4*, 1015. doi:10.3389/fpsyg.2013.01015
van Wijngaarden, S. J. (2001). Intelligibility of native and non-native Dutch speech. *Speech Communication, 35*(1-2), 103–113. doi:10.1016/S0167-6393(00)00098-4
Vigário, M., Butler, J., & Cruz, M. S. (2015). Phonologically constrained variability in L1 and L2 production and perception [Special Issue]. *Phonetica, 72*(2-3), 69–75. doi:10.1159/000441730
Werker, J. F., & Tees, R. C. (1984). Cross-language speech perception: Evidence for perceptual reorganization during the first year of life. *Infant Behavior and Development, 7*(1), 49–63. doi:10.1016/S0163-6383(84)80022-3

Research Article

Phonetica 2019;76:100–125
DOI: 10.1159/000494928

Received: April 3, 2017
Accepted: October 29, 2018
Published online: May 21, 2019

Gran Canarian Spanish Non-Continuant Voicing: Gradiency, Sex Differences and Perception

Karolina Broś Katarzyna Lipowska

University of Warsaw, Warszawa, Poland

Abstract

Background/Aims: This paper examines the process of postvocalic voicing in the Spanish of Gran Canaria from the point of view of language change. A perception-production study was designed to measure the extent of variation in speaker productions, explore the degree to which production is affected by perception and identify variables that can be considered markers of sound change in progress. ***Methods:*** 20 native speakers of the dialect were asked to repeat auditory input data containing voiceless non-continuants with and without voicing. ***Results:*** Input voicing has no effect on output pronunciations, but voicing is highly variable, with both phonetic and social factors involved. Most importantly, a clear lenition pattern was identified based on such indicators as consonant duration, intensity ratio, absence of burst and presence of formants, with the velar /k/ as the most affected segment. Furthermore, strong social implications were identified: voicing degrees and rates depend both on the level of education and on the gender of the speaker. ***Conclusion:*** The results of the study suggest that the interplay of external and internal factors must be investigated more thoroughly to better address the question of phonetic variation and phonologisation of contrasts in the context of language change.

Introduction

The process of voiceless non-continuant (/p t k tʃ/) voicing has been described as an incipient sound change taking place in several Romance languages, including Corsican (Gurevich, 2004), Sardinian (Jones, 1997; Loporcaro, 2009), Italian (Dal-

This paper is an extended version of a poster presentation given at OCP 14 in Düsseldorf, February 2017, which provided data on the voicing variable from Gran Canarian speaker productions. We would like to thank the conference participants for all their useful comments and suggestions concerning further developments of the study.

KARGER

E-Mail karger@karger.com
www.karger.com/pho

Karolina Broś
Dzielna 78/5
PL–01-029 Warszawa (Poland)
E-Mail k.bros@uw.edu.pl

cher, 2008; Hualde & Nadeu, 2011) and Spanish.[1] In the latter language, several dialects seem to present varying degrees of intervocalic fortis stop voicing, from Peninsular varieties (e.g., Lewis, 2001; Martínez Celdrán, 2009; Torreblanca, 1976; Torreira & Ernestus, 2011) to Central America and the Caribbean (Guitart, 1978; Quilis, 1993). The phenomenon seems to be gaining ground not only in dialects that are traditionally considered non-conservative (e.g., varieties spoken in Andalusia or the Caribbean), but also in other parts of the Spanish-speaking world. It is, however, still not stable or categorical, and depends on a variety of factors. Many studies point to both inter- and intraspeaker variation in the production of /p t k/. In a small study encompassing 4 speakers of Spanish from Barcelona, Machuca Ayuso (1997) reports fortis stop voicing rates ranging from 34.9 to 64.6% in spontaneous speech depending on the person. The degrees of voicing differed, and some stops were produced as approximants. In a fairly recent study, Hualde et al. (2011) examined spontaneous and scripted productions of 20 native Spanish speakers from Majorca. The analysis revealed that the voicing of /p t k/ in intervocalic position is fairly frequent (35.7% in the spontaneous speech corpus, compared to only 3.6% in read sentences). Hualde et al. also report substantial interspeaker variation. Interestingly, the voiced outputs of /p t k/ were usually realised as continuants in their study, which makes them very similar to phonological /b d g/ that are approximantised in the same environment. Such a state of affairs raises questions about the phonological merger of contrasts, hence other phonetic variables pointing e.g. to differences in the degree of constriction are worth measuring. Hualde et al.'s analysis demonstrated that there is a difference in consonant duration and constriction degree depending on the voicing of /p t k/ and compared to underlying voiced segments. No phonological merger, however, was confirmed due to the discrepancies between underlyingly voiced and voiceless sounds when compared by place of articulation, although the velar stop showed the greatest level of lenition. Herrera Santana (1997) and Lewis (2001) point to speech style as an important factor governing differences in the rates of voicing, with formal styles showing the change /p t k/ –> [b d g] only sporadically. Hualde et al. (2011) mention yet another variable – stress. In their study of /p t k/ voicing, there is a clear difference in the frequency of [b d g] depending on the type of syllable the stop belongs to. Finally, the place of articulation is said to play a role in the degree of weakening (e.g., Torreira & Ernestus, 2011). It is worth mentioning at this point that in our treatment of postvocalic voicing we look at the process from the perspective of language change and, more specifically, lenition/weakening. The latter phenomenon encompasses all those phonetic and phonological processes that gradually lead to segment elision, hence voicing, spirantisation/approximantisation, debuccalisation and related sound changes. These usually involve sound shortening, greater aperture (weaker constriction) and greater coarticulation (loss or overlap of individual articulatory gestures, target undershoot) which are consequences or manifestations of greater articulatory ease (effort minimisation), higher speech rate, biomechanical limitations and lower speech precision. Various internal and external factors contribute to sound weakening, some of which are discussed later in this paper.

[1] As suggested by a reviewer, it should be noted that in the case of Spanish we are talking about a second round of voicing from the historical perspective (Latin roots with intervocalic voiceless stops underwent voicing centuries ago, e.g. vita -> vida [biða] "life"), whereas in the case of Corsican, Sardinian or Central-Southern Italian the original Latin voiceless stops show variable voicing now.

As for perception, there are not many studies that address this issue in connection with stop voicing and possible merger with /b d g/, although Romero et al. (2007), for instance, argue that native speakers can reliably distinguish between underlyingly voiced and voiceless stops regardless of their surface constriction, voicing and duration. The latter feature has been reported as an important indicator of both voicing and constriction (e.g. Lavoie, 2001; Parrell, 2011). Martínez Celdrán (1993) states that if the period of silence preceding the release of a voiceless stop is shortened, the consonant will be perceived as voiced. In another study, he shows that speakers are confused when hearing words with voiced stops in isolation but are able to reliably distinguish meanings when confronted with full phrases in a perception test (Martínez Celdrán, 2009). This suggests that voicing obscures comprehension and context is needed to get at the meaning of a word with a voiceless stop that underwent voicing. Interestingly, however, speakers of the same dialect (here: Murcia Spanish) were better at recognising words than speakers of a different variety (Barcelona).

Given the abundance of phonetic studies showing variation in the production of voiceless non-continuants, information concerning the way this is processed by listeners and how it affects their future outputs is key to understanding phonetic variability and sound change. The present study aims to address this issue by providing production data correlated with perception. More specifically, we did not aim to test the perception of voiced outputs in itself (as related to the recognition of lexical contrasts). Rather, we wanted to use audition as a variable that potentially affects speaker productions. Our goal was also to see which features govern the differences between underlying and surface sounds, and whether they affect listener behaviour. The dialect taken under scrutiny, i.e. Gran Canarian Spanish, has been described as abounding in various types of consonantal lenition, including high rates of intervocalic voicing (Herrera Santana, 1989; Marrero, 1986; Oftedal, 1985; Trujillo, 1980). As reported in the literature, speakers of Canary Islands Spanish freely produce voiced variants of [p t k] in spontaneous speech regardless of age, sex or social status, although Trujillo (1980) points out that the phenomenon is more widespread among speakers from rural areas. The social reach of voicing was confirmed recently by Broś (2016a) based on fieldwork recordings of natural speech produced by Gran Canarian natives of different ages. The process involves all voiceless non-continuants /p t k tʃ/ both inside words and across word boundaries in strictly postvocalic position. Some examples are provided below.[2]

[2] It must be noted that in the Spanish of the Canary Islands, the prepalatal (or palato-alveolar) affricate has been reported to be produced as a palatal affricate [cç] or even stop [c], hence its voiced counterpart would be [ɟʝ] or [ɟ]. Depending on the study, the symbols and names used for the Canarian sound differ, which is why we decided to use IPA symbols for clarity. In our data, as in e.g. Dorta (1997), the affricate is often pronounced differently from the standard palato-alveolar [tʃ] found in Peninsular Spanish. Since pronunciations differ from one speaker to another, and the exact place and manner of articulation are outside the scope of this paper, we decided to use the [tʃ] symbol to refer to it and describe it as a prepalatal affricate in accordance with the Spanish-speaking literature. It should be noted, however, that the periods of closure and frication vary across speakers and depending on the degree of voicing. In general, our data show that females tend to produce a standard [tʃ] with a definite frication period and no voicing, whereas males typically produce a tenser variant whose frication period is shorter or even absent, with higher rates of voicing. The latter sound would be best referred to as "adherent ch" described by Álvar and Quilis (1966), Dorta (1997) and others.

DOI: 10.1159/000494928

Examples of postvocalic voicing observed both inside words and across word boundaries:

a. /p/ de[b]artamento "department"
b. /t/ foné[d]ica "phonetics"
c. /k/ má[g]ina "machine"
d. /tʃ/ le[dʒ]e "milk"
e. /p/ yo [b]ienso "I think"
f. /t/ juntos y [d]al "together etc."
g. /k/ de [g]olombia "of Colombia"
h. /tʃ/ la [dʒ]ica "the girl"

Apart from intervocalic position, Broś (2016a) reports voicing before a liquid consonant (e.g., *la playa* [la.**b**lá.ja] "triple"), which seems to follow the trajectory of consonantal changes observed historically in the transition from Latin to Spanish (Lloyd, 1987). The phenomenon seems to be very frequent in spontaneous productions, although it does not affect all voiceless non-continuants in an utterance. As it is sensitive to pauses and speech rate, it can be deemed coarticulatory. Furthermore, speech style is believed to be a major factor influencing the rate of voicing. As reported by Herrera Santana (1997), Gran Canarian speakers asked to read lists of words containing intervocalic voiceless non-continuants voice them only rarely. As for other accounts, Trujillo (1980) deems the process an important change in progress, and Oftedal (1985) alludes to the historical shift in the phonological categories of Spanish stops as a parallel of what can be observed today. The reported propensity for voicing and the simultaneous variation in production point to the phonetic nature of the process. Nevertheless, interaction with phonological phenomena puts such an interpretation into question. As demonstrated by Broś (2016b), voicing is blocked by the deletion of a preceding sound even though it should occur given the phonetic context. For instance, the phrase *pensar tonterías* "to think silly things" is realised [pensá**t**ontería], i.e. with a voiceless [t] although the final *r* of the word *pensar* is deleted. The same speaker regularly produces voicing in underlyingly postvocalic contexts. This makes Gran Canarian voicing phonological or at least allows the phonetic component of the grammar to interpret information provided by the phonological structure.

Experiment

Given the above facts, we decided to check which parameters are important in the production of voicing in the Gran Canarian variety of Spanish. We wanted to see whether the rates of voicing in experimental conditions would be similar to the ones reported for spontaneous speech, and whether they depend on auditory input. The latter question is important from the point of view of perception and interpretation of speech by listeners. In other words, we were interested in the speakers' response to variability. It is a generally accepted fact that in speech processing, production is mediated by the perception and interpretation of the auditory signal, as well as other factors. Because the phonetic output of a given individual may vary depending on a series of variables, such as speech rate, style, context or interlocutor, this may affect how listeners perceive and (re)produce speech during communication. Moreover, as argued by Ohala (1981, 1990), the intention of the interlocutor and the way it is discerned by the listener in communication is one of the main factors influencing sound change. Speakers base their interpretations on actual pronunciations, hence they may interpret unintended articulator movements resulting from a faster speech rate, overlap and undershoot as intended targets. In this way, sporadic or accidental phonetic changes generalise, and sound change can be observed in a given speech community.

Table 1. Participant data

Participant	Age, years	Sex	Education	Participant	Age, years	Sex	Education
1H	24	M	University degree	11M	37	F	University degree
2H	26	M	Student	12M	24	F	University degree
3M	23	F	Student	13H	25	M	University degree
4M	27	F	High school	14M	25	F	University degree
5H	24	M	High school	15M	23	F	Student
6H	24	M	University degree	16M	22	F	High school
7H	25	M	University degree	17M	23	F	High school
8H	25	M	High school	18H	27	M	Student
9H	33	M	University degree	19M	17	F	High school
10H	25	M	University degree	20H	32	M	University degree

To see how differences in production are processed and reproduced by speakers, a perception-production study was conducted among native speakers of Gran Canarian Spanish. Audio stimuli in two versions – voiced and voiceless – were designed to mimic interaction in communicative situations. Such an experiment design was chosen to elicit productions based on the semantic meaning and the mental lexicon rather than orthography. Avoiding read speech favours more natural, everyday articulation, and the use of voiced stimuli as inputs should promote voicing rather than standardisation.[3] We expected that experiment participants would voice the target sounds, and that they would be quite consistent in their productions. We did not expect much interspeaker variability (most speakers were in their twenties and from similar backgrounds); however, we wanted to see whether there are differences in production depending on the place of articulation.

Methods

Participants

20 speakers of Gran Canarian Spanish (11 males and 9 females) aged 17–37 years (mean 25.5, SD 4.3) participated in the experiment. They were recruited in the region of Gáldar, Las Palmas de Gran Canaria. 10 of them had a full university degree, 4 were university students at the time of the recording, and 6 had secondary school education. Table 1 presents participant details.

Stimuli

The stimuli used in the experiment consisted of 49 phrases containing contexts for postvocalic voicing across a word boundary. Each sentence was structured in the same way: *He comprado cinco* "I have bought five," followed by a noun phrase (NP), e.g. *panes de millo* "corn

[3] An important reason for avoiding written language and using variable productions in the input is the fact that consonant weakening is stigmatised in the speech of the Canary Islands. This is subject to a heated debate, and various authors point to the rates of "correction," teaching Canarians Peninsular Spanish pronunciations in schools and diffusing them as a model in the media (Moreno Cabrera, 2000; Morera, 1990).

DOI: 10.1159/000494928

breads." The NP always consisted of a noun denoting a container or an object made of other objects/ingredients (the target word), followed by a prepositional phrase introduced by *de* "of," or sometimes *con* "with" or *en* "in." The left-hand environment was always the vowel [o], whereas the target segment was [p], [t], [k] or [tʃ], where postvocalic voicing is expected. There were 10 different NPs per sound plus 3 different NPs per [pr], [tr] and [kr] sound combinations. The latter set was used to control for right-hand environments other than vowels given that the analysed process is postvocalic rather than strictly intervocalic. Stop plus liquid clusters are frequent in Spanish. The combination [tʃr], however, does not exist, hence it was not taken into account in the study. As for the choice of words for the stimuli, they were all checked with a native speaker of the dialect. It is difficult to reliably establish their frequencies given that no Canary Islands Spanish corpus for oral speech exists. Most words are frequently used in general Spanish (Spanish Corpus CREA and RAE dictionary, www.rae.es). Some were local words generally known by the population and listed in the dictionary of Canarian terms (http://www.academiacanarialengua.org), e.g. *traba*, general Spanish *pinza* "clothespin," *chincheta* "pushpin" or *trucha* – a Canarian sweet typically eaten during Christmas. The full list of stimuli is presented in the Appendix. Sentence length was limited given the nature of the task. We wanted to make sure that each participant would be able to repeat the stimuli based on auditory perception only.

Given that the experiment was based on audio input data, the stimuli had to be prerecorded. A trained male native speaker from Gáldar produced two sets of phrases: with and without voicing.[4] The recordings were made using a Zoom H4N digital recorder and a Shure SM10a headworn microphone. The same equipment was used during the experiment. The input phrases were subsequently cut and analysed in Praat (Boersma & Weenink, 2015) to ensure voicing where needed and control for pauses and overall recording quality. Fillers constructed in the same way as test sentences, with vowels, fricatives and voiced sounds instead of the target sounds were added. All audio files were then put together, and two randomised sets were created, 1 per 10 participants.

Procedure

The participants were asked to wear the microphone and headphones connected to a laptop computer and sit in front of a desk. Sound files with the stimuli were prepared to be played by the experimenter. The volume level was the same for all the participants. They confirmed that they could hear the recordings properly. After sitting down, they were instructed that upon hearing each sentence played by the experimenter they were to repeat it as naturally as possible. If they made a mistake during the repetition phase, they were asked to repeat the whole sentence. Subsequently, the experimenter would start the procedure, playing each stimulus in a predetermined order from a separate audio file. It was assumed that each sentence could be heard only once. Although some speakers had to repeat some of the sentences twice due to slips of the tongue, it was not necessary to repeat the recording in any of the cases.

[4] The choice of the speaker was based on the fact that he was a definite frequent voicer. The gender of the speaker was not of our concern given that according to previous studies, voicing is quite consistent in all speakers, both males and females of different ages. Since the study was not primarily focused on accommodation or speaker discrimination, both genders were listening to one speaker. As noted by a reviewer, however, this might have influenced the results given that participants may have voiced more readily if they were of the same gender as the recorded speaker. We see this as a limitation of the study. Nevertheless, according to our results, input voicing neither influenced output voicing in general nor when the two genders are compared (neither males nor females accommodate to the speaker from the recording). The only significant effect was observed in the case of the prepalatal (see the Results section). The question of covert prestige or lack thereof and the corresponding behaviour of the participants should not be ignored, however. Unfortunately, although two of the females were the heaviest voicers among the participants of the study and hence seemed to have an idiolectal tendency to voice stops everywhere, we cannot rule out the possibility that at least some of the other females reacted to the input in a negative way and adjusted their pronunciations because the speaker was male (e.g., to differentiate themselves from him).

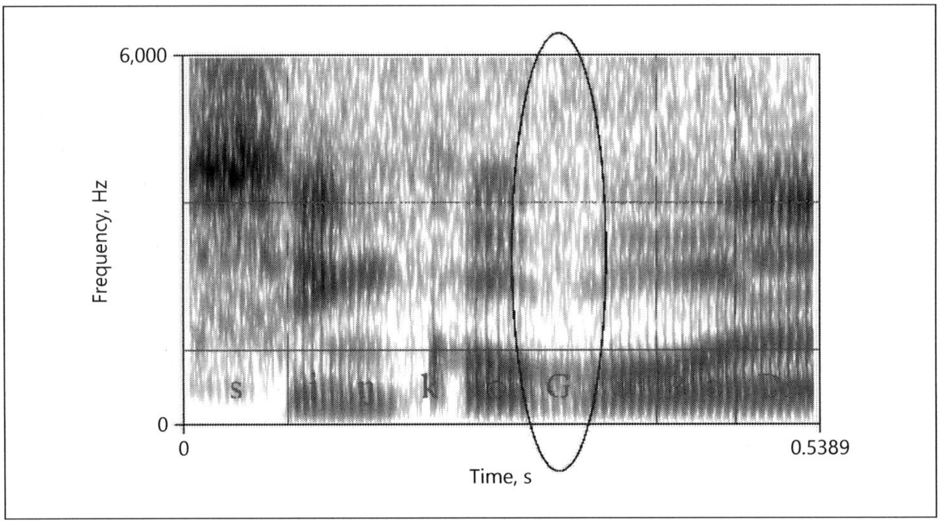

Color version available online

Fig. 1. Spectrogram of the words *cinco cubos* "five bins" presenting slight approximantisation (some formants visible). Capital letters designate approximantised consonants.

Measurements

To determine whether voicing is systematic and whether it follows the pattern produced in the stimuli, the voicing bar, glottal pulses and the voice report generated by Praat (Boersma and Weenink, 2015) were examined for each of the target sounds manually, in the course of visual inspection of the spectrograms. Given the voicing trail of the preceding vowel and other phonetic considerations, a rigorous classification of the sounds was made. It was decided that a given stop is voiced if pulses are present throughout the sound. It was annotated as partially voiced if more than 50% of the sound duration showed voicing on the spectrogram. Otherwise, the sound was deemed not voiced. Partial and full voicing were measured and then analysed both separately and when pooled together for comparison. Following the general analysis of output voicing and its relation to the input voicing of the target sounds, further measurements were made to assess the degree of weakening.

The degree of aperture was assessed indirectly via visual inspection of the spectrograms. It was assumed that weakened consonants present less constriction as they tend to be more vowel-like in terms of muscle tenseness and proximity of the articulators (Vennemann, 1988; Zec, 1995). One of the acoustic measurements of such changes in manner is the presence of *formants*, hence variable approximantisation of input stops. In our data, we observed variation in this respect, which was annotated and submitted for statistical analysis per sound and in correlation with other variables. More specifically, when some but not all formants were present on the spectrogram, the sound was annotated as weakly approximantised (marked with the digit 1, see Fig. 1). When a clear formant pattern was seen and the sound looked more like a sonorant, with only very slight formant transitions with respect to the flanking vowels, it was annotated as strongly approximantised (marked with 2, see Fig. 2).

Intensity ratio and *intensity difference* were calculated as two other indicators of the degree of constriction of a given consonant. For this purpose, two measurements were taken: the minimum intensity of the target sound and the maximum intensity of the following vowel (in decibels). The first measurement divided by the second gives the intensity ratio whose value is expected to be closer to 1 with a higher degree of lenition (Carrasco et al., 2012). The lower the ratio (i.e. the greater the difference between the two intensity values), the more consonantal a given sound. Voiced non-continuants should have a higher intensity ratio when compared to the

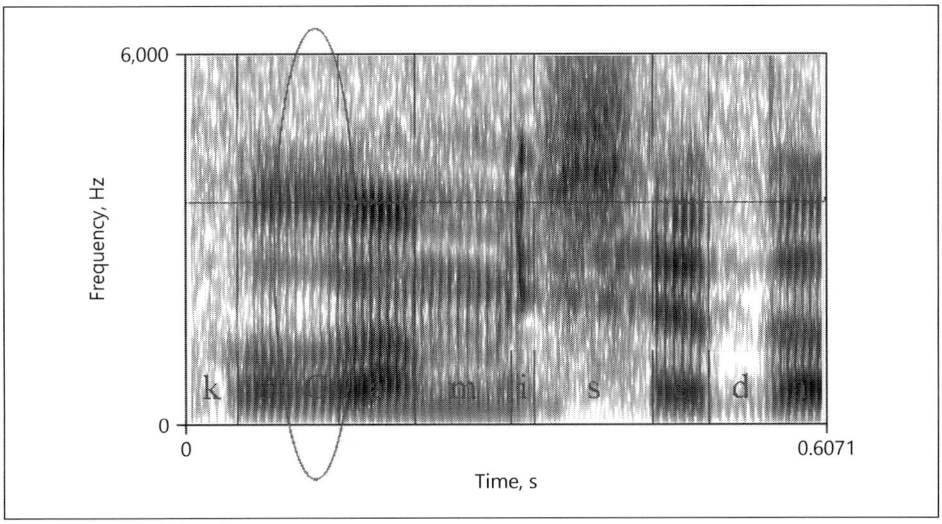

Fig. 2. Spectrogram presenting the word *camisetas* "T-shirts" presenting strong approximantisation (all formants visible). Capital letters designate approximantised consonants.

unvoiced ones. Consonants showing stronger lenition markers, such as formants, should also have a higher intensity ratio. This measure, alongside similar variables, has been used in several studies on consonant weakening (e.g., Colantoni & Marinescu, 2010; Hualde et al., 2010; Ortega-Llebaria, 2004). Carrasco et al. (2012) have shown that there are significant differences in the intensity ratio between voiceless stops and stops voiced by lenition in Spanish. Some studies (e.g., Hualde & Nadeu, 2011) used another intensity-related index, namely intensity difference measured by subtracting one of the above-mentioned values from the other (consonant from vowel). We decided to use it in our study for comparative purposes as several papers have shown that using more than one intensity measure helps to provide more comprehensive information on the degree of lenition (e.g., Hualde et al., 2011; Hualde & Nadeu, 2011). In this case, the more constricted the consonant, the greater the intensity difference, hence we expect lower values of this index in the case of lenition.[5] In /pr/, /tr/ and /kr/ clusters, intensity measurements were made on the vowel following the /r/. This follows the methodology used by Hualde & Nadeu (2011) who measure relative intensity with the omission of a glide or liquid following the stop in VCGV/VCLV sequences in Rome Italian.

Burst absence rate was deemed important in lenition processes by e.g. Dalcher (2008). Since only maximally constricted consonants produce bursts, complete closure must be maintained long enough (at least 20–30 ms) to ensure sufficient air pressure build-up and then abrupt release (Shadle, 1997). In our data, we observed that bursts are not necessarily produced and may or may not correlate with the voicing of a given stop, hence the variable was included in the statistical analysis.

Finally, *constriction duration* was measured as another lenition indicator. In accordance with the literature, constriction is expected to decrease proportionately to the degree of weakening. According to Dalcher (2008), weaker variants are progressively shorter, and the percentage of voicing

[5] It is worth mentioning that Parrell (2010) compared the two indices used in this study with one more variable (maximum velocity), concluding that intensity ratio is the most reliable index as it shows the strongest correlation with articulatory data. Although it is unclear how the intensity ratio should be calculated based on Parrell's presentation, the methodology is stated explicitly by other authors, e.g. Carrasco et al. (2012, p. 156) whose statistical analysis is based on "the intensity ratio resulting from dividing the two values [in decibels] (min/max) at the two points."

DOI: 10.1159/000494928

Table 2. Realisations of /p t k tʃ/: number of words produced with/without voicing, with formants and without burst in the corpus and their respective proportions with respect to all consonants or clusters of a given type expressed in percent

Consonant	Voiceless	Partially voiced	Fully voiced	Formants	No burst
/p/	178	73	138	99	132
	45.76%	18.77%	35.48%	25.45%	33.93%
/t/	214	60	114	101	72
	55.15%	15.46%	29.38%	26.03	18.56%
/k/	199	40	154	184	244
	50.64%	10.18%	39.19%	46.70%	61.93%
/tʃ/	232	80	80	43	n/a
	59.18%	20.41	20.41%	10.97%	n/a
/pr/	66	23	28	31	32
	56.41%	19.66%	23.93%	26.50%	27.35%
/tr/	77	16	26	33	18
	64.71%	13.45%	21.85%	27.73%	15.13%
/kr/	60	9	44	50	48
	53.10%	7.96%	38.94%	44.25%	42.48%
Total	1,026	301	584	541	548

Grand total: 1,911 sounds (389 /p/, 388 /t/, 393 /k/, 392 /tʃ/, 117 /pr/, 119 /tr/, 113 /kr/).

depends on constriction duration. At the same time, however, she has shown that *relative consonant duration* appears to be a more reliable indicator of weakening than relative constriction duration in Florentine Italian. The two variables were taken into account in the statistical analysis. The former is measured as a ratio of total consonant duration (constriction + release/positive voice onset time) to total VCV sequence duration, while the latter is a ratio of constriction duration to total VCV sequence. In this way, the values can be compared across speakers. VCRV sequences had to be excluded from the data for this measurement given that no reliable comparison can be made as to any of the ratios (both depend on the length of the sequence in milliseconds). Also, some items had to be excluded from length and/or constriction measurements when they were produced as strong approximants. In such cases, the presence of formants made it impossible to reliably establish the boundary between the consonant and the flanking vowels.

In all measurements, items with pauses/hesitations were excluded from statistical analysis to provide the most reliable results possible. The total number of analysed sounds was 1,911, and in the case of analyses excluding the affricate (burst, formants, constriction) – 1,520 (1,171 excl. clusters). A general summary of the results per sound (raw data) is presented in Table 2.

Statistical Methods

The analyses presented below were all conducted in R 3.4.2 (R Core Team, 2017) with the package lmer4 (Bates, Maechler, Bolker, & Walker, 2018). Linear mixed effects regression models were estimated using the function lmer(); for mixed effects binary logistic regression models, the glmer() function was employed with binomial family and the logit link function. Plots and estimated means were generated using the Effect() function from the effects package (Fox, 2003). In each model there were random intercepts for individual speakers. Full random structure was not employed due to problems with model convergence within 10,000 iterations. Dependent variables and fixed effects are described separately for each model.

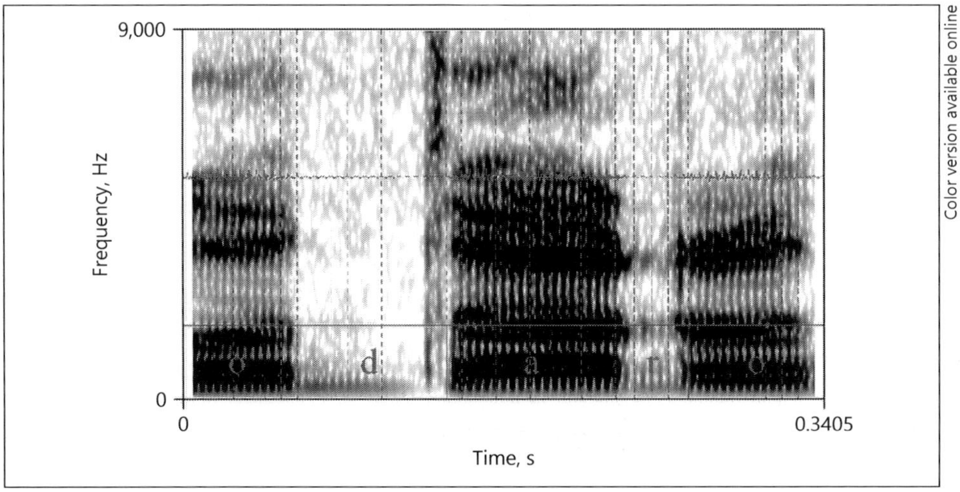

Fig. 3. Spectrogram of the word *tarros* "jars" presenting partial voicing (speaker 5H).

Results

The study revealed that voicing is not as consistent as assumed prior to running the experiment. The average voicing rate, including full and partial voicing, was only 44.9%, although significant rates of voicing were observed outside the target word across speakers. What is more, the process was not categorical. In accordance with our methodology, full voicing was produced in 29.5% of the cases, and partial voicing in 15.4%. Spectrograms presented in Figures 3 and 4 show partial and complete voicing, respectively. It is worth noting, however, that there was substantial interspeaker variation in our data. Four speakers voiced non-continuants in more than 75% of the cases. Another 4 had rates of voicing lower than 5%, including one who did not voice any of the target sounds. Seven people were in the mid-range (25–75%). Also, there were substantial differences between male and female productions. The collected data suggest that females voice less often (29.3% compared to 59.0% in males).

A mixed effects binary logistic regression was performed with absence/presence of voicing as the dependent variable, and gender, education (university level or high school), input (voiced/unvoiced) and sound (/tʃ/, /k/, /p/, /t/) as independent variables. The effects of sound ($F = 8.0$, df = 3, $p < 0.0001$)[6], gender ($F = 4.9$, df = 1, $p < 0.05$), sound × gender ($F = 6.0$, df = 3, $p < 0.001$) and sound × education ($F = 3.7$, df = 3, $p < 0.0001$) were significant (Fig. 5). The estimated voicing probability was highest for /p/ (52.7%), followed by /k/ (44.7%), /t/ (36.0%) and /tʃ/ (27.3%). This stands in contrast with findings concerning lenition tendencies in Romance (Recasens, 2002) and voicing in other Spanish dialects. Hualde et al. (2011), for instance, report /k/ as the most frequently voiced sound in intervocalic position. Interestingly,

6 The *p* for *F* values is given on the basis of a log likelihood test.

DOI: 10.1159/000494928

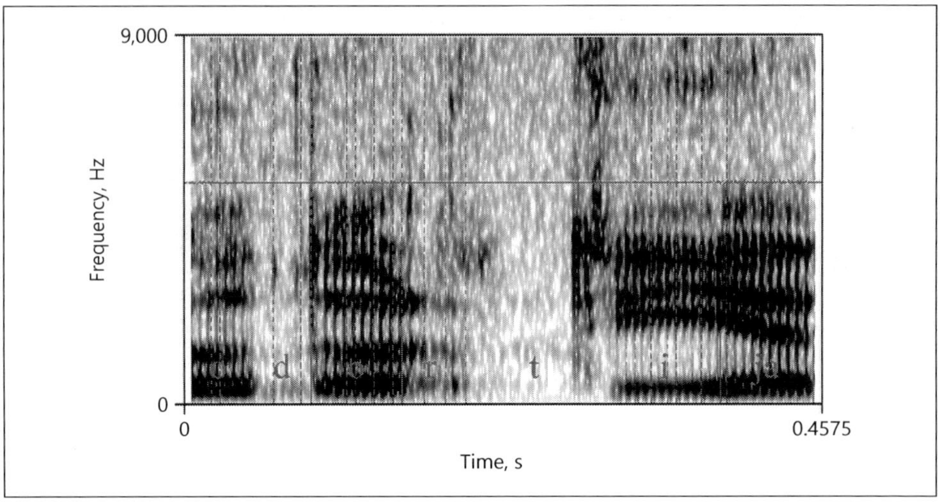

Color version available online

Fig. 4. Spectrogram of the word *tortillas* "omelettes" presenting full voicing (speaker 5H).

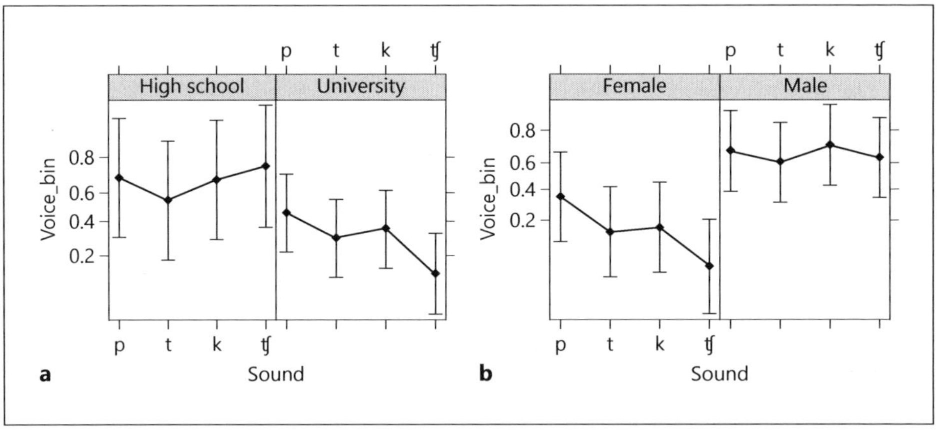

Color version available online

Fig. 5. Effect plots of sound × education (**a**) and gender × sound (**b**) interactions (mean predicted probabilities from the model).

according to our model, the differences in voicing between /p/ and /k/ were only significant in the case of university-educated women (see Table 3 for details). Furthermore, when the speaker was in high school, the voicing rate for /tʃ/ equalled 55.9% (on a par with the voicing rate for /p/, which was 57.1%), but for university-educated speakers this rate equalled 34.3% (the least of all examined sounds; Table 3). Also, female speakers tended to particularly "avoid" voicing /tʃ/ (raw data show a 21.0% voicing rate compared to 56.9% in males, and the difference was statistically significant). In order to further test the sociolinguistic hypothesis that different factors in-

Table 3. Binary logistic mixed model of voicing as a function of intervocalic sound (/p/, /t/, /k/, /tʃ/), participant gender and education with speaker as a random effect

Gender	Education	Voicing of /p/ compared to	Estimate	*z* value	*p* value
Female	high school	/k/	–0.66395	–1.771	0.07660
		/t/	–1.02120	–2.702	0.00689**
		/tʃ/	–0.71847	–1.847	0.06468
	university	/k/	–1.0526	–3.097	0.00195**
		/t/	–1.0993	–3.250	0.00115**
		/tʃ/	–2.7616	–6.279	3.40e-10***
Male	high school	/k/	0.44336	1.058	0.28995
		/t/	–0.31205	–0.777	0.43721
		/tʃ/	1.20026	2.510	0.01208*
	university	/k/	0.05501	0.197	0.84361
		/t/	–0.38975	–1.436	0.15111
		/tʃ/	–0.84295	–3.107	0.00189**
Sound	**Education**	**Voicing of /tʃ/ compared to**	**Estimate**	***z* value**	***p* value**
/tʃ/	high school	/p/	0.71942	1.850	0.064341
		/t/	–0.30204	–0.762	0.446329
		/k/	0.05524	0.140	0.888709
	university	/tʃ/	–3.02824	–2.824	0.004739**
		/p/	2.7618	6.283	3.33e-10***
		/t/	1.6625	3.761	0.000169***
		/k/	1.7092	3.848	0.000119***

/p/ is voiced significantly more often than /k/ in university-educated females. There are no significant differences in voicing of /p/ and /k/ in males. As for the /tʃ/, university-educated speakers voice it the least often, but there are no significant differences between the voicing of /tʃ/ and the voicing of other sounds in high school speakers. * $p \leq 0.05$, ** $p \leq 0.01$, *** $p \leq 0.001$.

fluence voicing in males and females, separate models for both genders were built. We found a statistically significant interaction of sound and education in the model for females ($F = 6.3$, $p < 0.0001$) but not in the model for males ($F = 1.6$, $p = 0.164$). In university-level females, the values for /t/ and /k/ are below 20%, while /tʃ/ voicing is not attested.

A slightly different binary logistic mixed model was employed to check the voicing in *r* clusters. Cluster presence ($F = 10.4$, $p < 0.01$) and sound ($F = 7.8$, $p < 0.0001$) were found to be significant. Additionally, when specific sounds were being compared on the basis of *z* values, gender was significant at $p < 0.05$, with females being less likely to voice sounds. Thus, the general conclusion is that clusters reduce the probability of voicing.

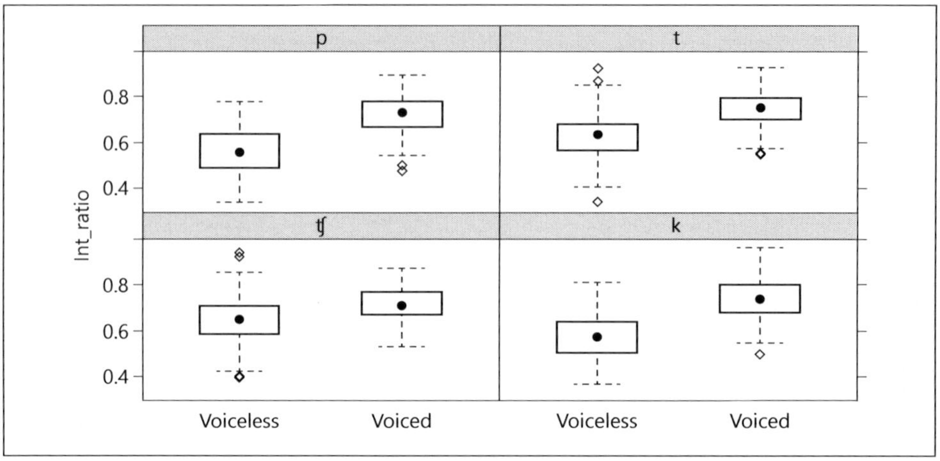

Color version available online

Fig. 6. Differences in the intensity ratio per sound depending on the voicing.

As for the influence of the input voicing on participant productions, no significant effect was found for all speakers taken together or for males/females separately, which means that output voicing was not affected by what speakers heard in the recording. Voicing was equally likely after hearing a voiced and a voiceless input (the distributions of voiced and voiceless sounds were around 50% each).

Other Variables

The overall *formant presence rate* was 28% (44% in males and 19% in females). The sound which presented formants most often is /k/ (46% of all outputs and 81% of voiced outputs, compared to 26% for both /p/ and /t/). In order to test whether formants in /k/ occur significantly more often than in /p/ and /t/, we built a mixed effects binary logistic regression model in R. There were significant effects of sound ($F = 19.9$, df = 2, $p < 0.0001$), voicing ($F = 113.2$, df = 1, $p < 0.0001$) and gender ($F = 4.5$, df = 1, $p < 0.05$). A comparison between sounds showed that there is a significant difference in formant occurrence in /k/ compared to /p/ and /t/. It can be therefore said that /k/ is the least constricted in production based on this indicator.

The mean *intensity ratio* was 0.66 (0.69 in males, 0.64 in females). Voiced productions had a mean of 0.73 and voiceless ones 0.60. Those differences were tested using a linear mixed effects regression model. The fixed effects of sound, voicing and their interaction all reached statistical significance ($F = 24.1$, df = 3, $p < 0.0001$, $F = 356.3$, df = 1, $p < 0.0001$, and $F = 27.9$, df = 3, $p < 0.0001$, respectively). Not only is the intensity ratio higher in the case of voicing, but the effect voicing has on the intensity ratio differs depending on the sound. For instance, the differences between voiced and voiceless realisations are much greater for /k/ and /p/ than for /tʃ/. The differences in intensity ratio depending on voicing per sound are presented in the form of a boxplot graph in Figure 6. These results are in line with the expected lenition pattern: the more lenited the stop, the less constricted or tense it is, hence the intensity ratio increases. Furthermore, there is a correlation between intensity and the presence of formants. As

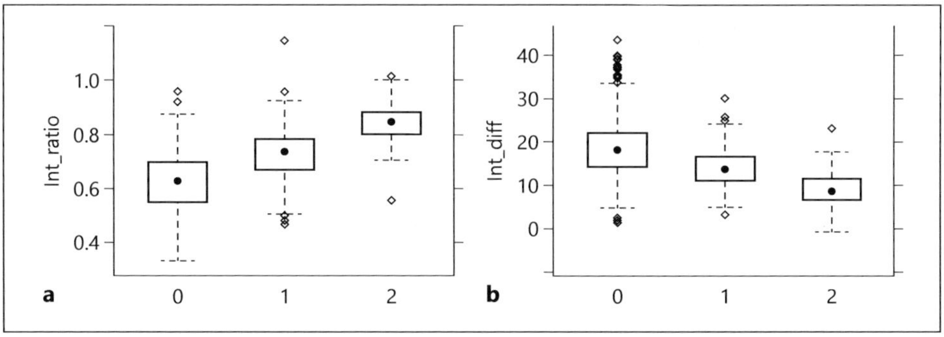

Fig. 7. Differences in the intensity ratio (**a**) and intensity difference (**b**) dependent on the presence/absence of formants. The boxplots are based on /p t k/ data. Slight to medium presence of formants is coded as 1, strong formants throughout the duration of the sound are marked by 2.

demonstrated in Figure 7, the intensity ratio rises gradually in proportion to formant presence (higher values in consonants presenting some formants, higher still when formants dominate the articulation of the sound throughout its duration).

The results for mean *intensity difference* are analogous to those obtained for intensity ratio (mean value 17.02, significant effects of sound, voicing and sound × voicing, all with a $p < 0.0001$). The only difference is that the relation between this variable and other lenition indicators goes in the opposite direction: the weaker the constriction, the smaller the intensity difference (Fig. 7). Thus, the role of intensity as a weakening marker was confirmed.

The *burst absence rate* was 36% across speakers, with a marked difference between the sexes (45% in males and 25% in females). There was also a substantial difference between voiceless (20.6%) and voiced or partially voiced (52.7%) productions. The sound that was most often produced without a burst was /k/ (58%), followed by /p/ (32%) and /t/ (18%). Again, a mixed effects binary logistic regression model was employed. Statistical significance was reached for sound ($F = 36.5$, df = 2, $p < 0.0001$), voicing ($F = 59.8$, df = 1, $p < 0.0001$), gender ($F = 3.8$, df = 1, $p = 0.051$) and sound × voicing ($F = 21.7$, df = 2, $p < 0.0001$). Apart from confirming the expected difference between sounds, burst was more likely in females (z value = 2.044, $p < 0.05$) and when there was no voicing (Fig. 8).

As for the *relative constriction duration*, the mean for voiced segments was 0.31, compared to 0.33 in voiceless consonants, with no gender difference. We found a significant effect of sound ($F = 36.8$, df = 2, $p < 0.0001$), burst ($F = 153.2$, df = 1, $p < 0.0001$), intensity ratio ($F = 175.9$, df = 1, $p < 0.0001$) and of the following interactions: gender × sound ($F = 5.3$, df = 1, $p < 0.05$), burst × voicing ($F = 31.7$, df = 1, $p < 0.0001$) and burst × voicing × formant ($F = 4.5$, df = 1, $p < 0.05$). According to our results, constriction duration was greatest when there was neither burst nor voicing; it then decreased when there was voicing, even more so if both voicing and burst were present. The shortest constriction duration was predicted in the case of presence of burst. This somewhat surprising pattern held except for /k/ in female speakers. For this reason, separate models were run for each sound to check for any other discrepancies. Again, the results were contrary to our expectations. For instance, the burst × voicing

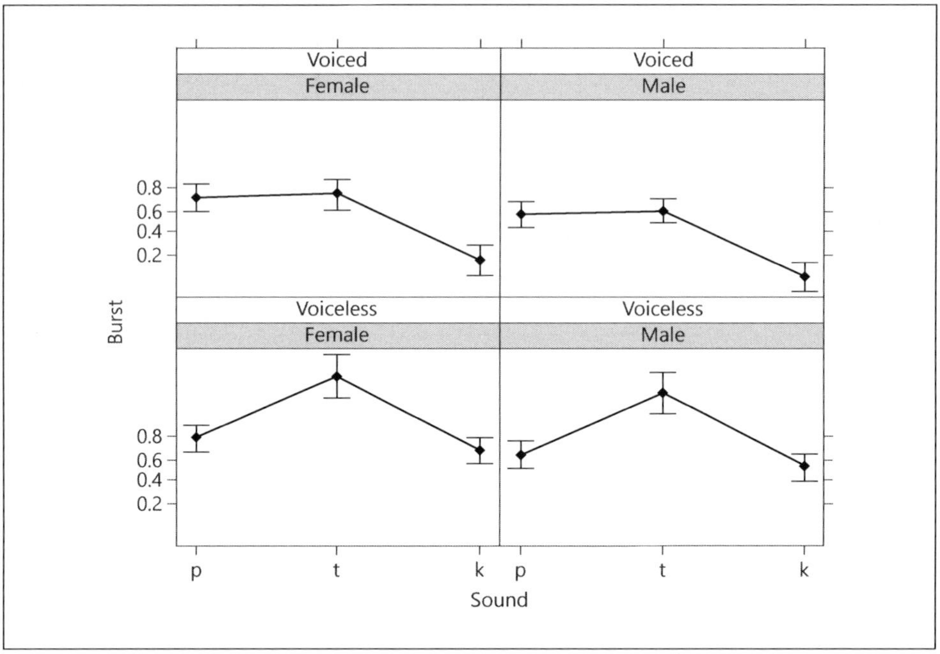

Fig. 8. Probability of burst occurrence by sound, voicing and participant gender.

interaction in /k/ looks as follows: in the absence of burst, constriction is greater without voicing, but the presence of burst *reverses* the effect of voicing, which instead of reducing constriction makes it longer. We attribute this effect to the fact that in the case of a stop produced with a burst, the constriction duration, measured from the offset of the previous vowel to the release burst, may be shorter than the constriction duration of a consonant without a burst, which is calculated from the offset of V1 to the onset of V2.[7] This variable is therefore not as reliable as total consonant duration in analysing data containing both released and unreleased stops. Similar conclusions were drawn by Dalcher (2008).

Thus, given the burst absence rate, we deem *relative consonant duration* more informative than relative constriction duration. Based on this variable, a clear lenition pathway from voiceless, through partially voiced to fully voiced stops was identified. As can be observed in Figure 9, the duration of the analysed sounds gradually decreases with voicing, regardless of the place of articulation. The average total duration for all consonants taken together was 64 ms in the case of voiced outputs (72 ms for partially voiced and 60 ms for fully voiced) and 87 ms in the case of unvoiced productions. The difference in the duration of all voiced consonants with respect to unvoiced tokens is 26.4%, but a clear leader can be discerned: whereas all individual sounds

[7] It should be noted that if no burst was visible but there was some energy or friction after the stop closure, this postclosure period was included in the constriction/consonant duration rather than considered to be a part of the onset of the following vowel.

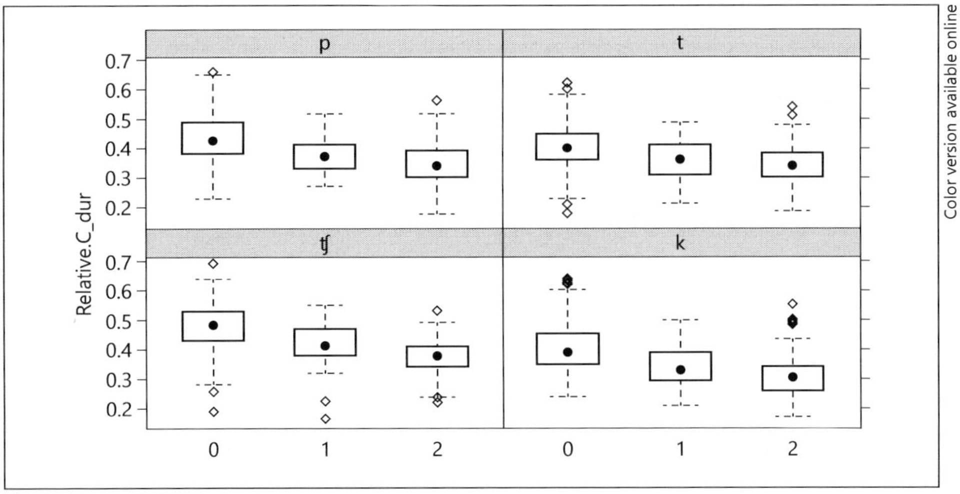

Fig. 9. Differences in relative consonant duration values per sound depending on the voicing. Voiceless productions are coded as 0, partially voiced tokens as 1 and fully voiced sounds as 2.

show a 23.7–23.8% decrease in duration whenever produced with voicing, /k/ duration drops by 30.2% (Table 4). Thus, despite the fact that it is not voiced the most often, /k/ is definitely shortened the most. The duration results can be treated as an indicator of lenition, similarly to other studies raising the question of how this variable correlates with sound weakening. Hualde and Nadeu (2011), for instance, find a clear difference in the durations of voiced and voiceless allophones of /p t k/ in Rome Italian. Their regression analyses point to a significant difference between the two sets of allophones with respect to one another, and with respect to underlying /b d g/, which makes duration an important correlate of lenition in stops. The differences in duration reported in this paper are similar to the results for /p t k/ and /b d g/ estimated by Parrell (2011). Other studies comparing the two classes of sounds in Spanish, however, report greater discrepancies (around 40–50 ms, with the shortest contrast in velars: approx. 18 ms; Lavoie, 2001). This may be due to the fact that underlyingly voiced stops typically realised as approximants were taken into account in that study, whereas we consider derived voiced stops only. There is reason to believe that durational differences between the latter two types of consonants would be observed in the dialect studied here in line with the general assumption that underlying differences take the form of different articulatory targets.

Additional Points

As reported by previous studies on lenition in Spanish, *stress* may also play a role. According to Hualde et al. (2011), the voicing of /p t k/ in spontaneous speech in Iberian Spanish increases from 23.5 to 39.4% in unstressed syllables. Similarly, in their investigation of Costa Rican Spanish lenition, Carrasco et al. (2012) found unstressed /b d g/ to be more lenited than stressed ones. In our study, 21 out of 49 target sounds were placed in an unstressed syllable (trisyllabic words). The rest occupy onsets of stressed syllables

Table 4. Mean durations of voiced and voiceless outputs in milliseconds

Consonant	Voiceless	Partially and fully voiced	Difference, ms	Difference, %
/p/	0.084	0.064	0.02	23.8
/t/	0.08	0.061	0.019	23.7
/k/	0.086	0.06	0.026	30.2
/tʃ/	0.101	0.077	0.024	23.7
Total	0.087	0.064	0.023	26.4

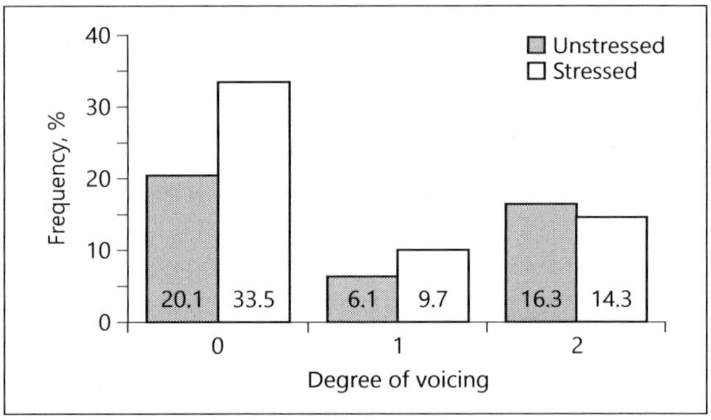

Fig. 10. Differences in the rates of voicing depending on stress. Partial voicing is coded as 1, full voicing corresponds to 2.

(in disyllabic words with penultimate stress). This makes it possible to compare the rates of voicing based on the stress variable. The analysis of the data reveals that there is an overall difference in the rates of voicing in the expected direction: there is more voicing in unstressed syllables (53% vs. 42% in stressed syllables). The results as per the degree of voicing are presented in Figure 10. A mixed effects binary logistic regression model was built with absence/presence of voicing as the dependent variable. Effects of stress ($F = 37.6$, df = 1, $p < 0.0001$) and sound ($F = 7.3$, df = 1, $p < 0.0001$) were significant. In the next step, we built a model for each sound separately to check which variables influence their respective voicing patterns (see summary in Table 5). In the model for /p/, only stress influenced the drop in the probability of voicing. In the models for /k/ and /t/, as expected, the presence of burst but absence of formants results in a smaller probability of voicing. Also, males were more likely to voice /k/ than women. In the model for /t/, we additionally found a significant effect of reduced probability of voicing when the syllable was stressed. /tʃ/ was more likely to be voiced in the absence of stress, when the speaker was male, in high school and had heard a voiced input.

Discussion

In our treatment of postvocalic voicing reported here, we are guided by the assumption that it is an instance of lenition, observable in real and apparent time and hence governed by the principles of language change. We are of the opinion that

DOI: 10.1159/000494928

pitch contour. An intonational break is also possible when repeating sentences after the speaker as the participants had to "make sure" they reproduce or remember the word from the audio stimulus correctly. In terms of prosodic boundaries, two interpretations of the stimuli are possible: *(He comprado cinco)*$_{\mathrm{PPh}}$ *(panes de millo)*$_{\mathrm{PPh}}$ and *(He comprado)*$_{\mathrm{PPh}}$ *(cinco panes de millo)*$_{\mathrm{PPh}}$ "I have bought five corn breads."[9] Our acoustic data seem to lend support to this prediction: the preliminary analysis of some phrase doublets produced by the same speaker with and without voicing shows slight differences in pitch (rise continuation vs. break). This might suggest different phrasing which can either favour or block voicing. Such an interpretation of the data can be explained by the theory of fine-grained phonetic modulation of speech at prosodic junctures (Cho, 2016), which assumes phonetic strengthening at domain boundaries (Fougeron & Keating, 1997). If the numeral is phrased separately, voicing will not be promoted, whereas if it is phrased together with the following NP, voicing may occur freely. In this way, we might also be able to explain the overall rate of voicing, which was much lower than expected.

Finally, we must mention the limitations of the study. First, as mentioned in footnote 6, the input data were produced by a male speaker, which may have influenced both male and female productions. Although the statistical analysis did not show any correlation between the voicing of the input and output productions, it is possible that at least some of the speakers spoke with a voicing frequency that does not reflect their everyday speech. The education factor is another possible limitation. The total number of speakers with high school education was 6 and only 2 out of 10 males were in this group, which means that the results showing male-female differences by education level should be treated with caution.

Conclusion

The data presented in this study disprove previous reports on intervocalic voicing in Gran Canaria, showing lower than expected rates of sonorisation, as well as substantial gradiency effects. Based on the results, we conclude that variation, which is a key feature of language change in progress exemplified here by Canary Islands Spanish, cannot be analysed strictly as a phonetic process dependent on internal factors. The trajectories of such sound change very much depend on social variables, such as gender, education and the implicit communicative choices of a given language user. The study is partially compatible with the biomechanical interpretation of gender differences. It also corroborated the Labovian gender paradox, especially the hypothesis that social status considerations are correlated with female behaviour in language production and women's resistance to non-prestigious changes compared to men. At the same time, variation in input voicing not only does not inhibit comprehension, but also seems not to affect output productions in a significant manner. The response to auditory input was positive only in the case of one sound and only when combined with other variables, hence we attribute the observed variation to both phonetic factors (coarticulation, prosody and speech rate) and socially grounded speaker pronunciation habits. Furthermore, although the comparison of

[9] The PPh abbreviation is used to refer to a minor phrase (phonological phrase), whose importance for Spanish has been argued for e.g. by Prieto (2007).

sound weakening by place of articulation showed differences attributable to the articulatory/aerodynamic constraints rather than phonology, all indices taken together, a clear set of weakening degree markers was identified. Voicing, intensity ratio/difference, burst absence rate, consonant duration and the presence of formants appear to be interrelated indicators of lenition regardless of the place of articulation. It further follows from this that the interplay between internal factors (prosody, stress, acoustic parameters) and external elements (gender, education) needs to be explored more deeply to help us understand the phonology of language change.

Appendix

List of Stimuli

1	He comprado 5 calcetines de Adidas	26	He comprado 5 paquetes de azúcar
2	He comprado 5 camisetas de algodón	27	He comprado 5 pares de zapatos
3	He comprado 5 chanclas de piscina	28	He comprado 5 piensos de perro
4	He comprado 5 chicles de fresa	29	He comprado 5 piezas de fruta
5	He comprado 5 chinchetas de colores	30	He comprado 5 pimientos de padrón
6	He comprado 5 chochos de yema	31	He comprado 5 piruletas de coca cola
7	He comprado 5 chocolates connueces	32	He comprado 5 postres de crema
8	He comprado 5 chocos en salsa	33	He comprado 5 prendas de ropa
9	He comprado 5 chorizos de Teror	34	He comprado 5 prensas hidráulicas
10	He comprado 5 chuletas de cerdo	35	He comprado 5 pruebas de embarazo
11	He comprado 5 chupas de bebé	36	He comprado 5 puerros de la granja
12	He comprado 5 chupitos de vodka	37	He comprado 5 tapas de tuétano
13	He comprado 5 colgadores de toallas	38	He comprado 5 tarros de garbanzos
14	He comprado 5 copas de vino	39	He comprado 5 tartas de queso
15	He comprado 5 crepes de nutella	40	He comprado 5 teléfonos de Nokia
16	He comprado 5 crocantes de almendra	41	He comprado 5 tenedores de plástico
17	He comprado 5 croquetas de atún	42	He comprado 5 terrones de azúcar
18	He comprado 5 cuadernos de notas	43	He comprado 5 tiendas de campaña
19	He comprado 5 cuadros de pintura	44	He comprado 5 tornillos de hierro
20	He comprado 5 cuartos de queso	45	He comprado 5 tortillas de papa
21	He comprado 5 cubos de basura	46	He comprado 5 tostadas de queso
22	He comprado 5 cuerdas de escalada	47	He comprado 5 trabas de la ropa
23	He comprado 5 kilos de tomates	48	He comprado 5 trozos de tarta
24	He comprado 5 panes de millo	49	He comprado 5 truchas de cabello
25	He comprado 5 pantalones de lana		

Acknowledgements

We would like to thank the anonymous reviewers of *Phonetica* and the Editors of this Special Issue for all their suggestions and constructive criticism, which definitely helped improve the quality of the paper.

Statement of Ethics

The subjects of this study have given their written informed consent, and the study was performed in accordance with the protocol and the common practice concerning linguistics research at the University of Warsaw.

Disclosure Statement

The authors have no conflicts of interest to declare.

Funding Sources

The study was partially funded by the University of Warsaw Foundation (05/2015).

Author Contributions

Each of the authors has made a substantial contribution to the conception and design of the submitted work, and participated in drafting, revising and approving the final version of the manuscript. In particular, the study was conducted and preprocessed phonetically in Spain. A detailed statistical analysis was then conducted in Poland. The two authors jointly interpreted the results, prepared the paper for publication and accept full responsibility for the contents thereof.

References

Álvar, M., & Quilis, A. (1966). Datos acústicos y geográficos sobre la ch adherente de Canarias. *Anuario de Estudios Atlanticos, 12*(1), 337–343.

Bates, D., Maechler, M., Bolker, B., & Walker, S. (2018). lme4: *Linear mixed-effects models using Eigen and S4. R package version 1.1–17.* Retrieved from https://cran.r-project.org/web/packages/lme4/lme4.pdf.

Boersma, P., & Weenink, D. (2015). Praat: doing phonetics by computer [Computer program]. Retrieved 10 February 2016 from http://www.praat.org/.

Broś, K. (2016a). Between phonology and morphosyntax: voicing and spirantization in the Spanish of Gran Canaria. In E. Cyran & J. Szpyra (Eds.), *Phonology, its faces and interfaces* (pp. 173–200). Frankfurt am Main: Peter Lang.

Broś, K. (2016b). Stratum junctures and counterfeeding: Against the current formulation of cyclicity in Stratal OT. In C. Hammerly & B. Prickett (Eds.), *Proceedings of the Forty-Sixth Annual Meeting of the North East Linguistic Society*. 157–170.

Cabrera Abreu, M., & Vizcaíno Ortega, F. (2010). Canarian Spanish Intonation. In P. Prieto & P. Roseano (Eds.), *Transcription of Intonation of the Spanish Language* (pp. 87–122). Munich: Lincom Europa.

Carrasco, P., Hualde, J. I., & Simonet, M. (2012). Dialectal differences in Spanish voiced obstruent allophony: Costa Rican versus Iberian Spanish. *Phonetica, 69*(3), 149–179. doi:10.1159/000345199

Cho, T. (2016). Prosodic boundary strengthening in the phonetics-prosody interface. *Language and Linguistics Compass, 10*(3), 120–141. https://doi.org/10.1111/lnc3.12178

Colantoni, L., & Marinescu, I. (2010). The scope of stop weakening in Argentine Spanish. In Ortega-Llebaria (Ed.), *Selected proceedings 4th Conference on Laboratory Approaches to Spanish Phonology* (pp. 100–114). Sommerville, MA: Cascadilla Press.

Dalcher, Ch. (2008). Consonant weakening in Florentine Italian: A cross-disciplinary approach to gradient and variable sound change. *Language Variation and Change, 20*(02), 275–316. doi:10.1017/S0954394508000021

Dorta, J. (1997). Datos acústicos y percepcion de la /c/ adherente de Canarias y de la prepalatal castellana. In M. Almeida & J. Dorta (Eds.), *Contribuciones al estudio de la lingüística hispánica (Homenaje al profesor Ramón Trujillo)* (pp. 57–72). Barcelona: Montesinos.

Eckert, P. (1989). The whole woman: Sex and gender differences in variation. *Language Variation and Change, 1*(03), 245–267. doi:10.1017/S095439450000017X
Fougeron, C., & Keating, P. A. (1997). Articulatory strengthening at edges of prosodic domains. *The Journal of the Acoustical Society of America, 101*(6), 3728–3740. doi:10.1121/1.418332
Fox, J. (2003). Effect displays in R for generalised linear models. *Journal of Statistical Software, 8*(15), 1–27. Retrieved from http://www.jstatsoft.org/v08/i15/. doi:10.18637/jss.v008.i15
Guitart, J. (1978). Aspectos del consonantismo habanero: Reexamen descriptivo. *Boletín de la Academia Puertorriqueña de la Lengua Española, 7*, 95–114.
Gurevich, N. (2004). *Lenition and contrast: The functional consequences of certain phonetically conditioned sound changes*. New York, NY: Routledge.
Guy, G. (2003). Variationist approaches to phonological change. In B. Joseph & R. Janda (Eds.), *The Handbook of Historical Linguistics* (pp. 369–400). Oxford: Blackwell. doi:10.1002/9780470756393.ch8
Herrera Santana, J. (1989). Sonorizacióm de oclusivas sordas en Tenerife. In J. Dorta & J. Herrera (Eds.), *Tres estudios de Fonética* (pp. 111–121). La Laguna: Universidad de La Laguna.
Herrera Santana, J. (1997). Estudio acústico de /p, t, c, k/ y /b, d, y, g/ en Gran Canaria. In M. Almeida & J. Dorta (Eds.), *Contribuciones al estudio de la lingüística hispánica (Homenaje al profesor Ramón Trujillo)* (pp. 73–86). Barcelona: Montesinos.
Hualde, J. I., & Nadeu, M. (2011). Lenition and phonemic overlap in Rome Italian. *Phonetica, 68*(4), 215–242. doi:10.1159/000334303
Hualde, J.I., Nadeu, M. & M. Simonet (2010). Lenition and phonemic contrast in Majorcan Catalan. In S. Colina, A. Olarrea, & A. M. Carvalho (Eds.), *Romance linguistics 2009*. Selected papers from the 39th Linguistic Symposium on Romance Languages (pp. 63–79). Amsterdam, NL: Benjamins.
Hualde, J. I., Simonet, M., & Nadeu, M. (2011). Consonant lenition and phonological recategorization. *Laboratory Phonology, 2*(2), 301–329. doi:10.1515/labphon.2011.011
Hume, E., & Johnson, K. (2001). A model of the interplay of speech perception and phonology. In E. Hume & K. Johnson (Eds.), *The role of speech perception in phonology* (pp. 3–26). New York: Academic Press.
Jones, M. A. (1997). Sardinian. In M. Maiden & M. Parry (Eds.), *The dialects of Italy* (pp. 376–384). London: Routledge.
Labov, W. (1980). The social origin of sound change. In W. Labov (Ed.), *Locating Language in Time and Space* (pp. 251–266). New York: Academic Press.
Labov, W. (1990). The intersection of sex and social class in the course of linguistic change. *Language Variation and Change, 2*(02), 205–254. doi:10.1017/S0954394500000338
Labov, W. (1994). Principles of Linguistic Change.: Vol. 1. *Internal Factors*. Oxford: Basil Blackwell.
Labov, W. (2001). Principles of Linguistic Change.: Vol. 2. *External Factors*. Oxford: Blackwell.
Labov, W. (2006). A sociolinguistic perspective on sociophonetic research. *Journal of Phonetics, 34*(4), 500–515. doi:10.1016/j.wocn.2006.05.002
Lavoie, L. (2001). *Consonantal Strength: Phonological Patterns and Phonetic Manifestations*. New York: Routledge. doi:10.4324/9780203826423
Lewis A.M. (2001). *Weakening of intervocalic /ptk/ in two Spanish dialects: Toward the quantification of lenition processes* (PhD dissertation), University of Illinois at Urbana-Champaign.
Lloyd, P. M. (1987). *From Latin to Spanish*. Philadelphia, PA: American Philosophical Society.
Loporcaro, M. (2009). *Profilo linguistico dei dialetti italiani*. Bari: Laterza.
Lucero, J. C., & Koenig, L. L. (2005). Phonation thresholds as a function of laryngeal size in a two-mass model of the vocal folds. *The Journal of the Acoustical Society of America, 118*(5), 2798–2801. doi: 10.1121/1.2074987
Machuca Ayuso M.J. (1997). *Las obstruyentes no continuas del español: Relación entre las categorías fonéticas y fonológicas en habla espontánea* (PhD dissertation), Universitat Autònoma de Barcelona.
Marrero, V. (1986). *Fonética estática y fonética dinámica en el habla de las Islas Canarias* (PhD dissertation), Universidad Complutense de Madrid.
Martínez Celdrán, E. (1993). La percepción categorial de /b, p/ en espanol basada en las diferencias de duración. *Estudios De Fonética Experimental, 5*, 223–239.
Martínez Celdrán, E. (2009). Sonorización de las oclusivas sordas en una hablante murciana: Problemas que plantea. *Estudios De Fonética Experimental, 18*, 253–271.
Moreno Cabrera, J.C. (2000). *La dignidad e igualdad de las lenguas: Crítica de la discriminación lingüística*. Madrid: Alianza Editorial.
Morera, M. (1990). *Unidad y variedad del español de Canarias*. La Laguna: Universidad de La Laguna.
Nadeu, M., & Hualde, J. I. (2015). Biomechanically conditioned variation at the origin of diachronic intervocalic voicing. *Language and Speech, 58*(Pt 3), 351–370. doi:10.1177/0023830914554727

Oftedal, M. (1985). *Lenition in Celtic and in Insular Spanish*. Oslo, Norway: Universitetsforlaget Oslo.
Ohala, J. (1981). The listener as a source of sound change. In C. S. Masek, R. A. Hendrik, & M. F. Miller (Eds.), *Papers from the parasession on language and behavior* (pp. 178–203). Chicago, IL: Chicago Linguistics Society.
Ohala, J. (1983). The origin of sound patterns in vocal tract constraints. In P. F. MacNeilage (Ed.), *The production of speech* (pp. 189–216). New York, NY: Springer. doi:10.1007/978-1-4613-8202-7_9
Ohala, J. (1990). The phonetics and phonology of aspects of assimilation. In J. Kingston & M. Beckman (Eds.), *Papers in laboratory phonology 1* (pp. 258-275). Cambridge, UK: Cambridge University Press. doi:10.1017/CBO9780511627736.014
Ortega-Llebaria, M. (2004). Interplay between phonetic and inventory constraints in the degree of spirantization of voiced stops: Comparing intervocalic /b/ and intervocalic /g/ in Spanish and English. In T. Face (Ed.), *Laboratory approaches to Spanish phonetics and phonology* (pp. 237–254). The Hague: Mouton de Gruyter.
Parrell, B. (2010). Articulation from acoustics: Estimating constriction degree from the acoustic signal. *Journal of the American Acoustical Society, 128*(4), 2289. doi:10.1121/1.3508033
Parrell, B. (2011). Dynamical account of how /b, d, g/ differ from /p, t, k/ in Spanish: Evidence from labials. *Laboratory Phonology, 2*(2), 423–449. doi:10.1515/labphon.2011.016
Prieto, P. (2007). Phonological phrasing in Spanish. In F. Martínez-Gil & S. Colina (Eds.), *Optimality-Theoretic Studies in Spanish Phonology* (pp. 39–61). Amsterdam, Philadelphia: John Benjamins. doi.org/10.1075/la.99.03pri
Quilis, A. (1993). *Tratado de fonética y fonología españolas*. Madrid: Gredos.
R Core Team (2017). *R: A language and environment for statistical computing*. Vienna, Austria: R Foundation for Statistical Computing. Retrieved from https://www.R-project.org/.
Recasens, D. (2002). Weakening and strengthening in Romance revisited. *Rivista di Linguistica*, 14, 327–373.
Romero, J., Parrell, B. & M. Riera (2007). What distinguishes /p/, /t/, /k/ from /b/, /d/, /g/ in Spanish? Poster presented at *Phonetics and Phonology in Iberia*, Braga, Portugal.
Shadle, C. H. (1997). The aerodynamics of speech. In W. Hardcastle & J. Laver (Eds.), *The handbook of phonetic sciences* (pp. 39–80). Oxford: Blackwell.
Torreblanca, M. (1976). La sonorización de las oclusivas sordas en el habla toledana. *Boletín De La Real Academia Española, 56*, 117–146.
Torreira, F., & Ernestus, M. (2011). Realization of voiceless stops and vowels in conversational French and Spanish. *Laboratory Phonology, 2*(2), 321–354. doi:10.1515/labphon.2011.012
Trudgill, P. (1972). Sex, covert prestige and linguistic change in the Urban British English of Norwich. *Language in Society, 1*(2), 175–195. doi:10.1017/S0047404500000488
Trujillo, R. (1980). Sonorización de sordas en Canarias. *Anuario Letras, 18*, 247–254.
Vennemann, T. (1988). *Preference laws for syllable structure and the explanation of sound change*. Berlin: Mouton de Gruyter.
Zec, D. (1995). Sonority constraints on syllable structure. *Phonology, 12*(01), 85–129. doi:10.1017/S0952675700002396

Research Article

Phonetica 2019;76:126–141
DOI: 10.1159/000493981

Received: March 31, 2017
Accepted: September 21, 2018
Published online: May 21, 2019

Perceptual Consequences of Variability in Native and Non-Native Speech

Melissa M. Baese-Berk Tuuli H. Morrill

Department of Linguistics, University of Oregon, Eugene, OR, USA

Abstract

Background/Aims: Native speakers often have a difficult time understanding non-native speech, and this challenge is frequently attributed to a more variable signal. While theories and models of general speech perception are grounded in issues of variability, they rarely consider non-native speech. Here, we ask how a specific type of variability (speaking rate) impacts two measures of perception for both native and non-native speech. ***Methods:*** In the present study, one group of listeners transcribed speech, providing a measure of intelligibility. A second group of listeners rated how fluent the speaker was, providing a measure of fluency. ***Results:*** The results show that variability in speaking rate correlates with a non-native speaker's intelligibility. However, perceived fluency measures are not predicted by this variability measure. ***Conclusions:*** These results, taken with studies of the range of variability in non-native speech, suggest that variability in non-native speech is not a monolithic construct. Current theories and models of perception can be enhanced by examining non-native speech and how variability in that speech impacts perception.

Introduction

A fundamental challenge to human communication is determining a speaker's intended meaning from a variable phonetic signal. Usually, listeners are able to accomplish this task with relative ease. However, under some circumstances this challenge results in more difficult listening conditions. One typical case of a difficult listening situation is when a native speaker listens to a non-native speaker. Substantial prior research has demonstrated that non-native speech is more difficult to under-

Melissa M. Baese-Berk
Department of Linguistics
161 Straub Hall, 1290 University of Oregon
Eugene, OR 97403 (USA)
E-Mail mbaesebe@uoregon.edu

stand than native speech (e.g., Bent & Bradlow, 2003; Bradlow & Bent, 2008; Lane, 1963; Munro & Derwing, 1995a, 1995b; van Wijngaarden, 2001). However, the factors underlying this difficulty have been difficult to pin down.

One hypothesis several researchers have put forth is that non-native speech is more variable than native speech (e.g., Wade, Jongman, & Sereno, 2007), and that this variability can contribute to difficulty in understanding speech (e.g., Witteman, Weber, & McQueen, 2014). If this hypothesis is correct, an increase in variability could make it more difficult for a listener to abstract constants from the speech stream to interpret the signal. In the present study, we examine this hypothesis by examining perceptual challenges as a function of the amount of variability in the speech signal. Because our previous work has demonstrated differences in variability in speaking rates between native and non-native speakers (Baese-Berk & Morrill, 2015; Morrill, Baese-Berk, & Bradlow, 2016), we ask whether speech intelligibility and judgments on speech fluency correlate with variability in speaking rate for both native and non-native speakers.

Most models of speech perception suggest that a primary challenge to speech perception is overcoming the variability that is inherent in the speech signal (Cooper, Delattre, Liberman, Borst, & Gerstman, 1952; Liberman, Cooper, Shankweiler, & Studdert-Kennedy, 1967). That is, the bulk of the theoretical and modeling work underlying speech perception has investigated how a listener is able to abstract constants from a varying acoustic signal (e.g., Blumstein & Stevens, 1981; Stevens & Blumstein, 1981). In recent years, theoretical approaches have shifted their focus to exemplar models wherein variability is meaningful in the signal (e.g., Johnson, 2005; Pierrehumbert, 2001), but the challenge for a listener to abstract a constant phonological representation still remains. The introduction of one possible contributor to variability – speech produced by non-native speakers – expands this challenge. Here, we ask how variability within a non-native speaker's production impacts perception of their speech. That is, we examine variability in speaking rate produced by non-native speakers and examine how this affects their intelligibility and perceived fluency, as judged by native listeners.

Non-native speech differs from native speech in many ways, at both segmental and suprasegmental levels. For example, non-native speech is substantially slower than native speech (Guion, Flege, Liu, & Yeni-Komshian, 2000). However, the bulk of the substantial previous work has examined the average properties of non-native speech. While this work certainly contributes to our understanding of non-native speech in general, very few studies have directly examined variability in non-native speech. The small body of literature that has asked whether non-native speakers are more variable in speech production than native speakers has demonstrated mixed results. Some work suggests that non-native speakers are *more* variable than native speakers (e.g., vowel dispersion in English by native Spanish speakers: Wade et al., 2007; speaking rate during read speech: Baese-Berk & Morrill, 2015). This has led to the assumption that non-native speech is always more variable than native speech (see Romero-Rivas, Martin, & Costa, 2015, and Witteman et al., 2014, for such assertions). However, other studies have suggested that, on some measures, non-native speakers are actually *less* variable than native speakers (function word duration: Baker et al., 2011; speaking rate during conversational speech: Morrill et al., 2016; vowel dispersion and consonant realizations in Japanese spoken by native speakers of English and Mandarin: Vaughn, Baese-Berk, & Idemaru, in press). While these findings lead to many interesting questions about under what circumstances speakers might be more or less variable in their production, they also introduce questions as to how variability impacts perception.

It is clear that non-native speech presents substantial challenges to native listeners (e.g., Adank, Evans, Stuart-Smith, & Scott, 2009; Bradlow & Bent, 2008; Flege et al., 1997; Floccia, Goslin, Girard, & Konopczynski, 2006). However, in spite of these difficulties, native listeners are able to adapt quickly to non-native speech (Baese-Berk, Bradlow, & Wright, 2013; Bradlow & Bent, 2008; Sidaras, Alexander, & Nygaard, 2009). Further, non-native listeners demonstrate relatively good perception of non-native speech regardless of whether or not they share the speaker's native language (e.g., Bent & Bradlow, 2003). The source of the initial difficulty of understanding non-native speech has often been attributed to the fact that it differs from native speech and creates an "adverse listening condition" (see, e.g., Mattys, Davis, Bradlow, & Scott, 2012). However, pinning down the exact factors that contribute to the initial difficulty in perception and to the rapid and robust adaptation to non-native speech has been substantially more challenging. Therefore, it is unclear how a modulation of variability might impact the perception of non-native speech.

In the present study, we focus on a suprasegmental feature of non-native speech, namely, speaking rate. Prosody, including timing information affected by the speaking rate, is one of the fundamental factors that influences perception and listener judgments on accent in non-native speech (e.g., Anderson-Hsieh, Johnson, & Koehler, 1992). However, suprasegmental features are relatively understudied in non-native speech. While it is clear that non-native speech is slower overall than native speech (e.g., Guion et al., 2000), recent work has suggested that non-native speakers also differ from native speakers with regard to their *variability* in speaking rate over time. Baese-Berk & Morrill (2015) demonstrated that when reading, non-native speakers show more variability in their speaking rate from utterance to utterance than native speakers do. That is, in addition to being slower overall, non-native speakers are also more variable in their speaking rates when reading. This finding reflects some previous results (e.g., Wade et al., 2007) that suggest that non-native speech is more variable than native speech. If non-native speech is more variable than native speech, this may be an underlying cause of difficulty for listeners. That is, in highly variable speech, it may be more difficult for a listener to abstract constants that are needed for comprehension. However, additional evidence suggests that the issue of variability in non-native speech is not quite so clean-cut and may differ as a function of the task performed, among other factors.

Only two studies have directly examined variability in speaking rate in non-native speech, and the findings from these two studies are mixed. As previously mentioned, Baese-Berk & Morrill (2015) demonstrated more variability in speaking rate for non-native speakers than for native speakers when reading aloud. When the same speakers from this study were recorded in spontaneous speaking tasks (i.e., telling stories from pictures or answering an interviewer's questions), the pattern was reversed: the non-native speakers showed less variability in their speaking rate from utterance to utterance than did the native speakers (Morrill et al., 2016). That is, comparisons of variability in speaking rate between native and non-native speakers differ as a function of the speaking style or task. This finding suggests that variability in speaking rate may be driven by very different factors in native and non-native speech, and the types of variability may behave differently in native and non-native speech.

It is possible that the variability shown by non-native speakers is driven by difficulties in planning for speech production (e.g., articulatory, lexical, or syntactic planning), whereas variability in native speech may index a level of mastery in which the speaker manipulates the speaking rate for discourse or expressive purposes. In storytelling, a speaker could use changes in speaking rate to signal important information, direct a listener's attention to (or away from) a particular content, or communicate emotional or other extralinguistic information. If changes in speaking rate are meaningful, these changes may impact how a listener perceives continuous speech, and the degree to which they perceive a non-native speaker as intelligible or fluent.

The specific features of this variability are particularly problematic for many models of speech perception which focus on the issue of abstraction over variation to come to a constant representation. While several models of perception predict how non-native listeners will understand speech in their non-native language (L2) (e.g., Best, 1995; Flege, 1995), no current models of speech perception directly predict how native listeners perceive non-native speech produced in the listener's native language (L1). In addition, these models only deal with variability as it is found within well-defined distributions of language-specific phoneme inventories. While models of L2 acquisition of prosody do exist (see, e.g., Mennen, 2015), these models also do not deal directly with variability. One assumption of many models of speech perception is that variability is a fundamental challenge to listeners – that is, the more variable a speaker is, the more difficult it will be to understand that speech. If we extrapolate this assumption to instances of non-native speech, it follows that in cases where non-native speech is *less* variable than native speech, this speech should be less challenging to understand. However, there is no substantial evidence that this is true. Because most models of speech perception focus either on native listeners listening to other native speakers or on non-native listeners acquiring their non-native language, it is unclear what role variability plays in the perception of non-native speech by native listeners. In fact, most models of speech perception currently cannot account for (a) the difficulty in initial understanding of non-native speech and (b) how listeners are able to quickly and robustly adapt to non-native speech with exposure (e.g., Baese-Berk et al., 2013; Bradlow & Bent, 2008).

Therefore, in the present study, we ask how a native listener processes variability in the speech of native and non-native speakers. That is, how does measured variability impact how a listener perceives the speech they are hearing? By answering this question, we hope to identify issues that need to be addressed in the development of theoretical models of speech perception, not only for non-native speech, but more generally for incorporating the nature of variability into native speech perception as well.

We specifically examine two aspects of perception: (1) perceived fluency and (2) intelligibility. Here, we define intelligibility as how many words a listener can transcribe accurately from a stream of speech (see Methods for the specific operationalization used in this experiment). Fluency is measured with a subjective rating from a listener who is asked to rate fluency on a scale of 1–9. We specifically chose these two measures because we hypothesized that they may behave differently with regard to variability and because they are not strictly correlated within a speaker. That is, a speaker could be highly intelligible and still rather disfluent. In the present work, we predict that intelligibility will correlate with variability; however, we anticipate that fluency may interact with variability in a more nuanced way. We outline the possible outcomes, and what factors may drive each possible result.

Previous work has suggested that under some circumstances (e.g., read speech) non-native speech is more variable than native speech, and in other cases (e.g., spontaneous speech) it is less variable (Baese-Berk & Morrill, 2015; Morrill et al., 2016). This suggests that variability could be taken by the listener to be both a marker of fluency and a marker of disfluency. For example, if a speaker, when speaking spontaneously, is able to modulate their speaking rate to convey pragmatic information, this may signal high degrees of fluency. However, if a speaker, while reading, modulates their speaking rate as a function of planning difficulties, motor challenges, or simply increased cognitive load, this may signal lower degrees of fluency. Therefore, a listener may realize that changes in speaking rate are not a reliable indicator of fluency and may rely less on such a measure when listening to speech. If this is the case, one might expect a weaker, or non-existent, correlation between variability in speaking rate and perceived fluency. However, it is also possible that a listener may believe that variability simply signals a lack of fluency, and this variability may strongly correlate with perceived fluency. In this case, the correlation may be stronger for non-native speakers, who tend to be perceived as less fluent than native speakers. It is clear from previous work that, overall, speaking rate does correlate with language skills in some cases (e.g., Flipsen, 2002; Miller, Grosjean, & Lomanto, 1984; Yaruss & Conture, 1995). However, it is important to note here that we are not examining cases of appropriate or inappropriate speaking rates. Rather, we are examining how the speaking rate shifts from utterance to utterance. Therefore, it is possible that variability in speaking rate may interact with fluency in different ways than do simple differences in speaking rate.

On the other hand, intelligibility may interact with variability in a more straightforward way. In the present study, we examine read speech. Previous research has suggested that non-native speakers demonstrate more variability in speaking rate than native speakers when reading (Baese-Berk & Morrill, 2015). Therefore, it is possible that variability may correlate strongly with intelligibility. It is possible that increased variability makes understanding non-native speech more difficult. It is also possible that increased variability also correlates with other features of non-native speech that may work in concert to make non-native speech less intelligible. Although the present study cannot disentangle these two possibilities, we address both issues in the Discussion.

Below, we present the results of an experiment consisting of fluency and intelligibility tasks. We describe how perceived fluency and intelligibility interact with variability in speaking rate in read speech by native and non-native speakers.

Subjects and Methods

Materials

Our speaker stimuli in this experiment were drawn from the Archive of L1 and L2 Scripted and Spontaneous Transcripts and Recordings (ALLSSTAR; Bradlow et al., 2011). We examined the speech of native English ($n = 10$), native Korean ($n = 10$), and native Mandarin ($n = 10$) speakers reading aloud in English. Speakers of these languages were chosen as they form the largest populations of speakers in the database, allowing us to examine a range of speakers. Further, these populations allow for a direct comparison with several other studies using the ALLSSTAR and Wildcat Corpora and use the same populations of speakers (e.g., Baese-Berk & Morrill, 2015; Baker et al., 2011; Morrill et al., 2016; Van Engen et al., 2010). Specifically, we used recordings of the speakers' productions of *The North Wind and the Sun* paragraph.

Participants

For the fluency experiment, 56 native English speakers (aged 18–26 years; 34 female) from the University of Oregon participated in the fluency task. The participants reported no significant speech or hearing disorders and reported no significant contact with non-native speakers of English. Specifically, the participants were asked if they had family members or close friends who were non-native speakers of English. Of course, this does not mean the participants had no exposure to non-native speech, simply that none of the listeners had substantial exposure that might have swayed the results. Further, our participants did not report significant exposure to other languages generally speaking, beyond a year or two of language education in a classroom. However, as in the case of exposure to non-native speakers of English, our listeners were relatively homogeneous, not showing substantial differences in their language background.

In the intelligibility experiment, 180 native English speakers participated via Amazon Mechanical Turk. The participants reported no significant speech or hearing disorders and reported no significant contact with non-native speakers of English.

Procedure

The two sets of participants described above completed two separate tasks: fluency and intelligibility. Because speaking rate change, our measure of variability, was calculated over pairs of utterances, the participants in the fluency task were presented with pairs of consecutive utterances from *The North Wind and the Sun* in PsychoPy (Peirce, 2007). For example, participants heard utterance 1 and utterance 2 from speaker A (see Appendix). This allowed us to compare the measure of variability with a measure of fluency over the same stretch of speech. The pairs of utterances were randomized.

The participants were asked to rate the utterances on a scale of 1–9, with 1 being "extremely disfluent" and 9 being "extremely fluent." They heard each pair of utterances once before making this decision. They then pressed a button to advance to the next pair of utterances. Fluency ratings were normalized using *z*-scores. We report those data below.

For the intelligibility task, the participants were presented with individual phrases from *The North Wind and the Sun*. Each participant transcribed a total of 9 phrases (see Appendix). For each listener, each phrase was produced by a different talker. Because each participant heard each phrase in the passage only once, a total of 6 participants transcribed each utterance by each talker. Each utterance was scored for the number of correctly transcribed key words (henceforth "key words correct") and for the total number of correctly transcribed words (henceforth "total words correct"). Key words were defined as all content words and critical function words. Non-critical function words (e.g., "a" vs. "the") were not counted in this tally.

Quantifying Variability

The methods used in the current study follow the method of quantifying variability used in our previous studies (Baese-Berk & Morrill, 2015; Morrill et al., 2016). The critical components of this method are summarized briefly below. To quantify variability in speaking rate, we employed a measure termed "rate change." This rate change measure quantifies the difference in speaking rate between pairs of utterances, by subtracting the speaking rate of one utterance from the speaking rate of the other utterance. For example, if utterance A has a speaking rate of 4 syllables per second and utterance B has a speaking rate of 5 syllables per second, the rate change would be 1 syllable per second. Similarly, if utterance C has a speaking rate of 3 syllables per second, the rate change between utterance B and utterance C would be –2 syllables per second. This measure of rate change can be expressed either as a raw value, which indicates whether a speaker is speeding up or slowing down their speech between utterances, or an absolute value, which quantifies whether a speaker changes their speaking rate but does not quantify the directionality of the change. Both measures are useful for examining variability in speaking rate (see Baese-Berk & Morrill, 2015, for more details).

Non-native speakers show significantly more variability in both measures than do native speakers in read speech, including in the stimuli used in the present study. This method was chosen because our previous work had demonstrated that rate change is independent of speaking rate, meaning that non-native speakers are not penalized for simply speaking more slowly (see,

Table 1. Summary of the full mixed-effects model examining fluency as a function of rate change, nativeness, and language background

Factor	Estimate	Standard error	*t* value	χ^2	*p* value
Intercept	6.1418	0.3606	17.031	–	–
Nativeness	–5.3275	0.5684	–9.373	42.33	<0.001
Language background	0.4128	0.4922	0.839	1.06	0.3
Rate change	–0.2485	0.1535	–1.619	2.22	0.13
Nativeness × rate change	–0.2693	0.4269	–0.631	<1	0.5
Language background × rate change	0.7421	0.3822	1.942	<1	0.5

The χ^2 values represent the differences in likelihood ratios when comparing a model including this factor to a model that does not include the factor. The *p* values are derived from the χ^2 (degrees of freedom in all comparisons are 1).

e.g., Guion et al., 2000). Further, the method gives a local measurement (i.e., a separate value for each pair of utterances), rather than a more global measurement such as the variation in speaking rate over longer stretches of time. This allows us to more directly correlate variation with measures of fluency and intelligibility on a local level.

For the present study, utterances were defined as pause-bounded stretches of speech. Because this was read speech in a fairly simple form, the utterances were identical for all speakers and corresponded to the utterances described in the Appendix. Pairs refer to utterances spoken in a sequence; therefore, we were capturing variability in speaking rate from utterance to utterance. In the analyses below, we examine correlations between a perceptual measure (i.e., perceived fluency or intelligibility) and the absolute value of rate change.

Results

Fluency

To examine the relationship between rate change and perceived fluency, we constructed mixed-effects models with perceived fluency as the dependent variable. All analyses were run using R (R Core Team, 2015). We contrast-coded language background such that we could compare the native speakers to the non-native speakers and then compare the non-native speaker groups to one another. The fixed effects included in the model were rate change, nativeness (i.e., native English speakers vs. both the native Korean and native Mandarin speakers), language background (i.e., native Korean speakers vs. native Mandarin speakers), and the interactions between rate change and nativeness and rate change and language background. Random effects in the model were the maximal that allowed for model convergence and included random intercepts for listener, item, and speaker. A summary of the full model is found in Table 1. Figure 1 presents the average rate change and average fluency rating for each speaker. The lines show the best-fitting regression line for each language group.

In order to determine significance, we conducted model comparisons comparing the full model to models created by removing one factor or interaction at a time.

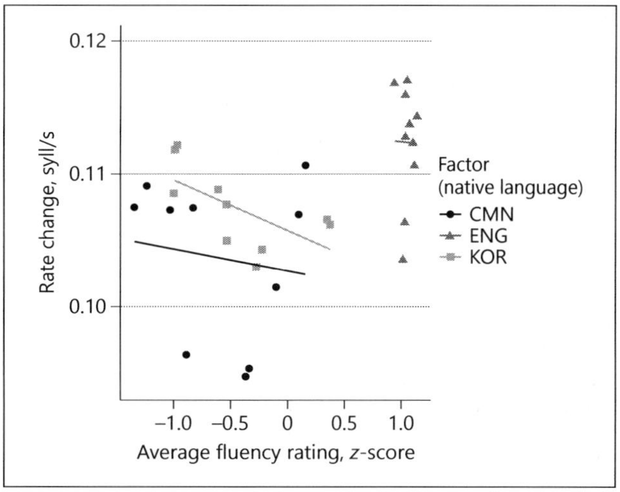

Fig. 1. Average fluency rating (*z*-score) and average rate change for speakers from three language backgrounds: Mandarin Chinese (CMN), English (ENG), and Korean (KOR). syll/s, syllables per second.

The inclusion of nativeness in the model significantly improved model fit ($\chi^2 = 42.33$, $p < 0.001$). As in Figure 1, it is clear that non-native speakers were rated as less fluent than the native speakers, overall. However, including language background in the model did not significantly improve model fit ($\chi^2 = 1.06$, $p > 0.3$), suggesting there was no significant difference in the perceived fluency of the native Korean and native Mandarin speakers. Similarly, the inclusion of rate change in the model did not significantly improve model fit ($\chi^2 = 2.22$, $p > 0.13$), suggesting that rate change did not significantly impact perceived fluency. Neither the interaction between rate change and nativeness nor the interaction between rate change and language background significantly improved model fit ($\chi^2 < 1$, $p > 0.5$).

Taken together, the results of the fluency test suggest that while non-native speakers are perceived as less fluent than native speakers, rate change, language background within the non-native group, and the interaction between those factors do not significantly predict perceived fluency.

Intelligibility

To examine the relationship between rate change and intelligibility, we constructed mixed models similar to those presented for the fluency test, using number of key words correct as the dependent variable. As in the models examining fluency, we contrast-coded language background. The fixed effects included in the model were the same as those in the fluency models and included rate change, nativeness (i.e., native English speakers vs. both the native Korean and native Mandarin speakers), language background (i.e., native Korean speakers vs. native Mandarin speakers), and the interactions between rate change and nativeness and rate change and language background. Random effects in the model were the maximal that allowed for model convergence and included random intercepts for listener, item, and speaker. A summary of this full model is presented in Table 2. Figure 2 presents the average rate change and average key words correct for each speaker. Figure 3 presents the average rate change and total words correct, demonstrating that the results show a

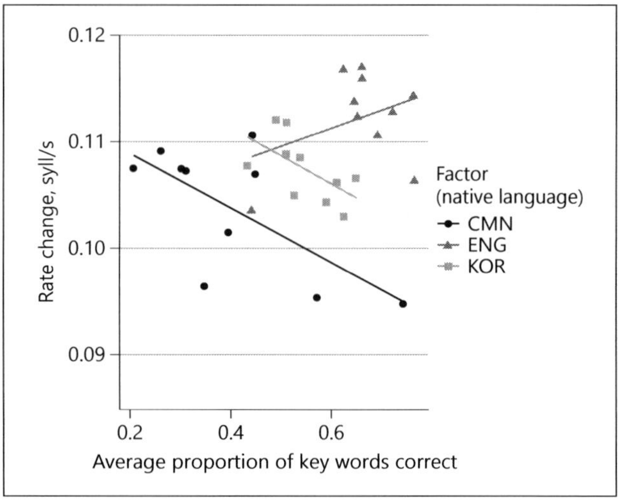

Fig. 2. Average proportion of key words correct and average rate change for speakers from three language backgrounds: Mandarin Chinese (CMN), English (ENG), and Korean (KOR). syll/s, syllables per second.

Table 2. Summary of the full mixed-effects model examining intelligibility as a function of rate change, nativeness, and language background

Factor	Estimate	Standard error	*t* value	χ^2	*p* value
Intercept	0.5957	0.03127	19.047	–	–
Nativeness	−0.27770	0.08812	−3.151	195.5	<0.001
Language background	0.51021	0.07193	7.093	40.47	<0.001
Rate change	−0.72687	0.27489	−2.644	187.99	<0.001
Nativeness × rate change	−0.78617	0.66874	−5.662	0.28	0.5
Language background × rate change	−0.26253	0.78562	−0.334	31.79	<0.001

The χ^2 values represent the differences in likelihood ratios when comparing a model including this factor to a model that does not include the factor. The *p* values are derived from the χ^2 (degrees of freedom in all comparisons are 1).

similar pattern regardless of whether the dependent variable is key words correct or total words correct.

As in the case of the fluency results presented above, we conducted model comparisons comparing the full model to models created by removing one factor or interaction at a time. As in the fluency test, the inclusion of nativeness (i.e., English speakers vs. the other two language groups) in the model significantly improved model fit ($\chi^2 = 195.5$, $p < 0.001$), suggesting that non-native speakers are less intelligible than native speakers. Interestingly, the inclusion of language background (i.e., Mandarin vs. Korean speakers) also significantly improved model fit ($\chi^2 = 40.47$, $p < 0.001$). Examining Figure 2, it is clear that native Mandarin speakers are less intelligible than native Korean speakers, which is reflected in the results of the model. Un-

DOI: 10.1159/000493981

Acknowledgements

This work was partially funded by a University of Oregon Faculty Research Award to M.M.B.-B.

Disclosure Statement

The authors declare no conflicts of interest with the current work.

Appendix

Utterance 1:	the North Wind and the Sun were disputing which was the stronger
Utterance 2:	when a traveler came along wrapped in a warm cloak
Utterance 3:	they agreed that the one who first succeeded in making the traveler take his cloak off should be considered stronger than the other
Utterance 4:	then the North Wind blew as hard as he could
Utterance 5:	but the more he blew, the more closely did the traveler fold his cloak around him
Utterance 6:	and at last the North Wind gave up the attempt
Utterance 7:	then the Sun shone out warmly
Utterance 8:	and immediately the traveler took off his cloak
Utterance 9:	and so the North Wind was obliged to confess that the Sun was the stronger of the two

References

Adank, P., Evans, B. G., Stuart-Smith, J., & Scott, S. K. (2009). Comprehension of familiar and unfamiliar native accents under adverse listening conditions. *Journal of Experimental Psychology. Human Perception and Performance, 35*(2), 520–529. https://doi.org/10.1037/a0013552

Anderson-Hsieh, J., Johnson, R., & Koehler, K. (1992). The relationship between native speaker judgments of nonnative pronunciation and deviance in segmentals, prosody, and syllable structure. *Language Learning, 42*(4), 529–555. https://doi.org/10.1111/j.1467-1770.1992.tb01043.x

Baese-Berk, M. M., & Morrill, T. H. (2015). Speaking rate consistency in native and non-native speakers of English. *The Journal of the Acoustical Society of America, 138*(3), EL223–EL228. https://doi.org/10.1121/1.4929622

Baese-Berk, M. M., Bradlow, A. R., & Wright, B. A. (2013). Accent-independent adaptation to foreign accented speech. *The Journal of the Acoustical Society of America, 133*(3), EL174–EL180. https://doi.org/10.1121/1.4789864

Baker, R. E., Baese-Berk, M., Bonnasse-Gahot, L., Kim, M., Van Engen, K. J., & Bradlow, A. R. (2011). Word durations in non-native English. *Journal of Phonetics, 39*(1), 1–17. https://doi.org/10.1016/j.wocn.2010.10.006

Bent, T., & Bradlow, A. R. (2003). The interlanguage speech intelligibility benefit. *The Journal of the Acoustical Society of America, 114*(3), 1600–1610. https://doi.org/10.1121/1.1603234

Bent, T., Baese-Berk, M., Borrie, S. A., & McKee, M. (2016). Individual differences in the perception of regional, nonnative, and disordered speech varieties. *The Journal of the Acoustical Society of America, 140*(5), 3775–3786. https://doi.org/10.1121/1.4966677

Best, C. T. (1995). A direct realist perspective on cross-language speech perception. In W. Strange (Ed.), *Speech perception and linguistic experience: Issues in cross-language research* (pp. 171–204). Timonium, MD: York Press.

Blumstein, S. E., & Stevens, K. N. (1981). Phonetic features and acoustic invariance in speech. *Cognition, 10*(1-3), 25–32. https://doi.org/10.1016/0010-0277(81)90021-4

Bradlow, A. R., Ackerman, L., Burchfield, L., Hesterberg, L., Luque, J., & Mok, K. (2011). Language- and talker-dependent variation in global features of native and non-native speech. In *Proceedings of the XVIIth International Congress of Phonetic Sciences* (pp. 356–359). Hong Kong, China: City University of Hong Kong.

Bradlow, A. R., & Bent, T. (2008). Perceptual adaptation to non-native speech. *Cognition, 106*(2), 707–729. https://doi.org/10.1016/j.cognition.2007.04.005

Cooper, F. S., Delattre, P. C., Liberman, A. M., Borst, J. M., & Gerstman, L. J. (1952). Some experiments on the perception of synthetic speech sounds. *The Journal of the Acoustical Society of America, 24*(6), 597–606. https://doi.org/10.1121/1.1906940

Flege, J. E. (1995). Second language speech learning: Theory, findings, and problems. In W. Strange (Ed.), *Speech perception and linguistic experience: Issues in cross-language research* (pp. 233–277). Timonium, MD: York Press.

Flege, J. E., Bohn, O. S., & Jang, S. (1997). Effects of experience on non-native speakers' production and perception of English vowels. *Journal of Phonetics, 25*(4), 437–470. https://doi.org/10.1006/jpho.1997.0052

Flipsen, P., Jr. (2002). Longitudinal changes in articulation rate and phonetic phrase length in children with speech delay. *Journal of Speech, Language, and Hearing Research: JSLHR, 45*(1), 100–110. https://doi.org/10.1044/1092-4388(2002/008)

Floccia, C., Goslin, J., Girard, F., & Konopczynski, G. (2006). Does a regional accent perturb speech processing? *Journal of Experimental Psychology. Human Perception and Performance, 32*(5), 1276–1293. https://doi.org/10.1037/0096-1523.32.5.1276

Guion, S. G., Flege, J. E., Liu, S. H., & Yeni-Komshian, G. H. (2000). Age of learning effects on the duration of sentences produced in a second language. *Applied Psycholinguistics, 21*(2), 205–228. https://doi.org/10.1017/S0142716400002034

Johnson, K. (2005). Speaker normalization in speech perception. In D. B. Pisoni & R. E. Remez (Eds.), *Handbook of speech perception* (pp. 363–389). Oxford: Blackwell. https://doi.org/10.1002/9780470757024.ch15

Klatt, D. H. (1979). Speech perception: A model of acoustic-phonetic analysis and lexical access. *Journal of Phonetics, 7*(3), 279–312.

Kraljic, T., Samuel, A. G., & Brennan, S. E. (2008). First impressions and last resorts: How listeners adjust to speaker variability. *Psychological Science, 19*(4), 332–338. https://doi.org/10.1111/j.1467-9280.2008.02090.x

Lane, H. (1963). Foreign accent and speech distortion. *The Journal of the Acoustical Society of America, 35*(4), 451–453. https://doi.org/10.1121/1.1918501

Laturnus, R. (2018). *Perceptual adaptation to non-native speech: The effects of bias, exposure, and input variation* (Unpublished doctoral dissertation), New York University, New York, NY.

Liberman, A. M., Cooper, F. S., Shankweiler, D. P., & Studdert-Kennedy, M. (1967). Perception of the speech code. *Psychological Review, 74*(6), 431–461. https://doi.org/10.1037/h0020279

Mattys, S. L., Davis, M. H., Bradlow, A. R., & Scott, S. K. (2012). Speech recognition in adverse conditions: A review. *Language and Cognitive Processes*, 27(7-8), 953–978. https://doi.org/10.1080/01690965.2012.705006

McLaughlin, D. J., Baese-Berk, M. M., Bent, T., Borrie, S. A., & Van Engen, K. J. (2018). Coping with adversity: Individual differences in the perception of noisy and accented speech. *Attention, Perception & Psychophysics, 80*(6), 1–12. https://doi.org/10.3758/s13414-018-1537-4

Mennen, I. (2015). Beyond segments: Towards a L2 intonation learning theory. In E. Delais-Roussarie, M. Avanzi, & S. Herment (Eds.), *Prosody and language in contact* (pp. 171–188). Berlin, Heidelberg: Springer. https://doi.org/10.1007/978-3-662-45168-7_9

Miller, J. L., Grosjean, F., & Lomanto, C. (1984). Articulation rate and its variability in spontaneous speech: A reanalysis and some implications. *Phonetica, 41*(4), 215–225. https://doi.org/10.1159/000261728

Morrill, T., Baese-Berk, M., & Bradlow, A. (2016). Speaking rate consistency and variability in spontaneous speech by native and non-native speakers of English. In *Proceedings of Speech Prosody* (pp. 1119–1123). Boston, MA: ISCA.

Munro, M. J., & Derwing, T. M. (1995a). Foreign accent, comprehensibility, and intelligibility in the speech of second language learners. *Language Learning, 45*(1), 73–97. https://doi.org/10.1111/j.1467-1770.1995.tb00963.x

Munro, M. J., & Derwing, T. M. (1995b). Processing time, accent, and comprehensibility in the perception of native and foreign-accented speech. *Language and Speech, 38*(Pt 3), 289–306. https://doi.org/10.1177/002383099503800305

Peirce, J. W. (2007). PsychoPy – Psychophysics software in Python. *Journal of Neuroscience Methods, 162*(1-2), 8–13. https://doi.org/10.1016/j.jneumeth.2006.11.017

Pierrehumbert, J. B. (2001). Exemplar dynamics: Word frequency, lenition and contrast. *Typological Studies in Language, 45*, 137–158. https://doi.org/10.1075/tsl.45.08pie

R Core Team. (2015). *R: A language and environment for statistical computing* [Internet]. Vienna, Austria: R Foundation for Statistical Computing. Available from: https://www.R-project.org/

Romero-Rivas, C., Martin, C. D., & Costa, A. (2015). Processing changes when listening to foreign-accented speech. *Frontiers in Human Neuroscience, 9*, 167. https://doi.org/10.3389/fnhum.2015.00167

Sidaras, S. K., Alexander, J. E., & Nygaard, L. C. (2009). Perceptual learning of systematic variation in Spanish-accented speech. *The Journal of the Acoustical Society of America, 125*(5), 3306–3316. https://doi.org/10.1121/1.3101452

Stevens, K. N., & Blumstein, S. E. (1981). The search for invariant acoustic correlates of phonetic features. In P. D. Eimas & J. L. Miller (Eds.), *Perspectives on the study of speech* (pp. 1–38). New Jersey: Erlbaum.

Van Engen, K. J., Baese-Berk, M., Baker, R. E., Choi, A., Kim, M., & Bradlow, A. R. (2010). The Wildcat Corpus of native- and foreign-accented English: Communicative efficiency across conversational dyads with varying language alignment profiles. *Language and Speech, 53*(Pt 4), 510–540. https://doi.org/10.1177/0023830910372495

van Wijngaarden, S. J. (2001). Intelligibility of native and non-native Dutch speech. *Speech Communication, 35*(1-2), 103–113. https://doi.org/10.1016/S0167-6393(00)00098-4

Vaughn, C., Baese-Berk, M., & Idemaru, K. (in press). Re-examining phonetic variability in native and non-native speech. *Phonetica*. https://doi.org/10.1159/000487269

Wade, T., Jongman, A., & Sereno, J. (2007). Effects of acoustic variability in the perceptual learning of non-native-accented speech sounds. *Phonetica, 64*(2-3), 122–144. https://doi.org/10.1159/000107913

Witteman, M. J., Weber, A., & McQueen, J. M. (2014). Tolerance for inconsistency in foreign-accented speech. *Psychonomic Bulletin & Review, 21*(2), 512–519. https://doi.org/10.3758/s13423-013-0519-8

Yaruss, J. S., & Conture, E. G. (1995). Mother and child speaking rates and utterance lengths in adjacent fluent utterances: Preliminary observations. *Journal of Fluency Disorders, 20*(3), 257–278. https://doi.org/10.1016/0094-730X(94)00013-J

Research Article

Phonetica 2019;76:142–162
DOI: 10.1159/000493983

Received: April 3, 2017
Accepted: September 21, 2018
Published online: May 21, 2019

Effects of Language Background on the Development of Sociolinguistic Awareness: The Perception of Accent Variation in Monolingual and Multilingual 5- to 7-Year-Old Children

Bronwen G. Evans[a] Gisela Tomé Lourido[a, b]

[a]Department of Speech, Hearing and Phonetic Sciences, University College London, London, UK; [b]School of Languages, Cultures and Societies, University of Leeds, Leeds, UK

Abstract

As a result of complex international migration patterns, listeners in large urban centres such as London, UK, likely encounter large amounts of variation in spoken language. However, although dealing with variation is crucial to communication, relatively little is known about how the ability to do this develops. Still less is known about how this might be affected by language background. The current study investigates whether early experience with variation, specifically growing up bilingually in London, affects accent categorization. Sixty children (30 monolingual, 30 bilingual) aged 5–7 years, were tested in their ability to comprehend and categorize talkers in 2 out of 3 accents: a home, unfamiliar regional and unfamiliar foreign-accented variety. All children demonstrated high, above-chance performance in the comprehension task, but language background significantly affected the children's ability to categorize talkers. Bilinguals were able to categorize talkers in all accent conditions, but although all children were able to understand the talkers, monolingual children were only able to categorize talkers in the home-foreign accent condition. Overall, the results are consistent with an approach in which gradient representations of accent variation emerge alongside an understanding of how variation is used meaningfully within a child's environment.

Introduction

Contact between speakers of different languages and accent backgrounds is a long-established phenomenon, but recently, national and international migration patterns have become more complex (e.g., Vertovec, 2007). In turn, this has created

KARGER

E-Mail karger@karger.com
www.karger.com/pho

Dr. Bronwen G. Evans
Department of Speech, Hearing and Phonetic Sciences,
University College London
Chandler House, 2, Wakefield Street
London, WC1N 1PF (UK)
E-Mail bronwen.evans@ucl.ac.uk

increasingly diverse multidialectal and multilingual communities, where people regularly come into contact not just with speakers of different regional accents, but also different languages (e.g., Cheshire et al., 2011). Children growing up in such diverse communities are thus likely to encounter speakers from a wide variety of backgrounds, native and non-native.

Urban centres like London are a prime example of this. For instance, in the recent 2011 UK census (Office for National Statistics, 2016), over a third of London's population reported that they had been born overseas (vs. 13% nationally) with over a fifth (22.1%) reporting that they spoke a main language other than English (vs. 8% nationally). This means that children growing up monolingually in the dominant language, English, are likely to be exposed not just to different local varieties such as Popular London and Multicultural London English (cf. Cheshire et al., 2011), but also to different national and international varieties of English, as well as so-called World Englishes (e.g., Indian English) and foreign-accented speech (i.e., speakers with English as a second language). Children growing up bilingually will likely be exposed to still more variation. They will be exposed to different accents of English (McCarthy et al., 2013; 2014), but also to additional language, accent and talker variation. These children may be raised as simultaneous bilinguals (i.e., exposed to both languages from birth, e.g., Mayr et al., 2015; Sundara & Scuttellaro, 2011) or as sequential bilinguals, who are initially exposed primarily to the family or community language, and then learn the host community language when they enter school or nursery education at around 3–4 years of age (Darcy & Krüger, 2012; McCarthy et al., 2014; Xu & Demuth, 2013). This means that on starting nursery or school education, sequential and possibly also simultaneous bilinguals are likely to have exposure not just to different languages and native-speaker varieties of English (e.g., through their monolingual peers, teachers) but also to non-native, i.e., foreign-accented varieties (e.g., through grandparents; see e.g., McCarthy et al., 2013).

Yet research on how young children's language development is affected by their immediate language environment and, in particular, the role that variation plays in language development, is scarce. What little we do know is largely limited to studies of children growing up monolingually in communities where children interact with talkers who speak similar varieties of the same language (e.g., Jones et al., 2017; Kinzler & DeJesus, 2013; Wagner et al., 2014; Weatherhead et al., 2016). In these studies, young children aged 5–6 years have been shown to be unable to distinguish an unfamiliar regional accent from their own accent, but to be able to tell apart their home accent from a foreign accent. One explanation for this finding is that, at least initially, children's representations are organized along a quantitatively defined continuum, anchored by their "home" accent. This is thought to form the core set of their experiences, with regional accents treated as intermediate between the home and second-language accent (Wagner et al., 2014). The current study aims to extend this work to examine the effects of early experience, specifically whether or not a child is raised monolingually or bilingually within a highly diverse speech community (London, UK), on the ability to extract and use patterns of variation in the speech signal to categorize home, unfamiliar regional (Yorkshire) and unfamiliar foreign-accented (Singapore English) talkers. Specifically, we aim to investigate whether children who are exposed to more variation develop a more detailed, gradient representation of variation within their community

The Role of Experience in the Processing of Accent Variation

Previous work on the role of variation in speech processing has focussed predominantly on its effect on comprehension. For adults, unfamiliar-accented speech has been shown to have an initial processing cost (see Cristia et al., 2012, for a review); listeners are slower to decide whether a word is real or not (Floccia et al., 2006), are more adversely affected by background noise (Clopper & Bradlow, 2008) and make more errors in deciding whether or not a sentence is true or false (Adank et al., 2009). However, with exposure they are able to rapidly adapt to these differences (Maye et al., 2008), perhaps after only 1–2 sentence lengths (Clarke & Garrett, 2004). Adaptation effects have also been found in long-term contact situations; native speakers of British English who have lived in the USA are better able to process American English speakers' flapped productions of medial-/t/, [ɾ], e.g., in *city*, where they would typically have [t] (Scott & Cutler, 1984). Likewise, Adank et al. (2009) found that Glaswegians living in Glasgow who were regularly exposed to Standard Southern British English (SSBE) were able to make true/false judgements on sentences presented in SSBE with the same speed and accuracy as when they were presented in their native accent, Glaswegian, but that SSBE speakers living in London with little or no experience of Glaswegian were slower and less accurate with Glaswegian than SSBE.

However, the mechanism by which listeners are able to adapt remains unclear. Some have interpreted these results as suggesting that speech processing is highly flexible and that, with enough exposure, listeners are able to adjust their category representations with relative ease (e.g., Johnson, 1997). In contrast, others have argued that abstract, underlying categories have limited flexibility, and that instead, short-term adaptation effects arise as a result of top-down changes, e.g., becoming better able to map variants to one's own underlying categories (Iverson & Evans, 2009). For example, in a series of experiments, Shaw et al. (2018) investigated whether Australian listeners' categorization of their own vowels, along with those in 4 unfamiliar varieties of English, was facilitated by short-term multi-talker exposure. In contrast to previous studies, (e.g., Maye et al., 2008; Norris et al., 2003), the results demonstrated that listeners' performance on the categorization task was unaffected by exposure. Based on these findings, the authors hypothesize that listeners' representations can tolerate a good range of natural variation (cf. Evans & Iverson, 2007), and that this ability to "hear through" variation in order to recover the phonological category may be what is fundamental to maintaining perceptual flexibility.

Evidence from studies investigating children's perception of accented speech provides further support for this view. Nathan et al. (1998) tested monolingual English children, aged 4–7 years, born and raised in London, in their ability to recognize words produced in their own (London) and a Glaswegian accent, as measured in a paradigm in which they had to repeat and then define a word. As expected, all children found it harder to understand words in the unfamiliar Glaswegian accent, but there were also developmental differences in how children responded; older children were more accurate in defining the word and also gave significantly more phonological responses (i.e., repeated the word in their own accent) than did younger children, who instead tended to respond using the Glaswegian pronunciation. For the majority (61%) of these phonetic responses, children were also unable to define the word, indicating that they had not achieved lexical access. One possibility is that this shift from phonetic to phonological processing enabled older children to better map the incoming signal to their own, underlying stored representations. That is, with more experience of varia-

DOI: 10.1159/000493983

tion, older children were able to retrieve phonological and lexical information, i.e., achieve phonological constancy (Best et al., 2009; Nathan et al., 1998).

Accent Variation, Categorization and the Role of Experience

Language acquisition must, however, also involve distinguishing the linguistic from the social, i.e., understanding how a given variable is used meaningfully within a speech community. For example, adults make inferences about people based on the speech signal (e.g., Bayard et al., 2001; Preston, 1989), and these judgements can change their basic categorization of speech stimuli (Strand, 1999). Indeed, listeners' stereotypes about gender, as activated by the faces and voices of speakers, have been shown to alter the perception of a phonetic continuum such that a female face or voice causes the perceived boundary between /s/-/ʃ/ to shift up in frequency, whilst a male face or voice causes the boundary to shift down in frequency (Strand, 1999). Moreover, faces rated by independent judges as highly masculine or feminine have been shown to lead to greater changes than those rated as less masculine and feminine (Strand, 1999). Similar effects have been found for regional accent. Niedzielski (1999) showed that evoking a specific region by writing the name of a place on a response sheet, caused listeners who were familiar with the accent spoken there to categorize vowels according to the way in which they are produced in that regional accent (cf., Hay & Drager, 2010). Together, these findings indicate that socially constructed beliefs play a role in basic phonological categorization, and that social information affects how listeners perceive and categorize speech sounds (see Campbell-Kibler, 2016, for an excellent discussion of how social and linguistic information might be integrated in models of cognitive processing).

Including a role for socially constrained information in speech processing is consistent with an exemplar-based model of speech perception (e.g., Goldinger, 1997; Hawkins & Smith, 2001; Johnson, 1997). This approach argues that linguistic knowledge is based on detailed memory representations which simultaneously encode non-linguistic (e.g., who was speaking, in what situation) and linguistic (e.g., segmental features, voice quality, pitch) information (Docherty & Foulkes, 2014; Munson, 2010). Any abstract representations are thought to arise from common factors amongst the many memories a speaker-listener has (Hawkins, 2003), such that knowledge of phonetic and social variation derives from the clustering of similar forms, with knowledge of socially structured variation emerging where linguistic and social differences coincide (Docherty & Foulkes, 2014). Such a model supports a probabilistic approach to phonetic learning and categorization (Foulkes & Docherty, 2006); as a speaker-listener builds up a library of traces, he/she develops sensitivity to the statistical properties of the usage of different variables, i.e., different patterns of usage. It is from this awareness that abstract, phonological categories emerge (Pierrehumbert, 2003; Pierrehumbert & Gross, 2003). Thus, a listener who has experience with different British accents may be able to recognize northern-accented speech by mapping it onto similar stored exemplars produced by northern talkers and recognize southern-accented speech by mapping it onto similar stored exemplars produced by southern talkers. Specifically, being able to make hypotheses about where a speaker is from may enable a listener to predict which phonetic variants are likely to be used and, consequently, enable them to tune in to the talker more easily. For example, hearing tokens of BATH produced with the northern English [æ] may "prime" a listener not only to expect more tokens of BATH words produced with this vowel, but also to expect to hear other northern English variants, such as STRUT and FOOT words produced with the same vowel, [ʊ], even if they have not yet heard tokens of this variable. Likewise, by adjusting their expectations of how /s/-/ʃ/ will be produced according to whether or not listeners think they are listening to a male or female talker (Strand, 1999), listeners may be better able to map the incoming signal to their own, underlying abstract representations of these categories.

In terms of child language development, we might hypothesize that as children build up a bank of detailed representations through experience with speech that is rich in language-specific and potentially talker-specific phonetic detail, there is a developmental progression from specific to more abstract knowledge of sound structure (Munson, 2010). Acquisition might therefore in part be driven by frequency of exposure, with those differences that are both phonetically large and experienced more frequently (e.g., male-female differences based on pitch range) emerging earlier than those which are more arbitrary (e.g., ethnicity, social class) and which might be experienced relatively rarely within the speech community (see Foulkes & Docherty, 2006).

Recent studies investigating children's ability to detect accent-related differences between talkers lend experimental support to this view. For example, children aged 5–6 years are able to group talkers according to their accent, but only when the phonological differences between the accents are maximized (Girard et al., 2008; Wagner et al., 2014). That is, they are able to successfully categorize native versus foreign-accented talkers but are unable to reliably tell apart talkers of their own accent and another unfamiliar regional accent where the acoustic-phonetic differences between the talkers are relatively small. They also understand that a speaker's accent can index where someone lives (Weatherhead et al., 2016) and can make meaningful social associations with a talker based on his/her accent. For example, they can link familiar (e.g., a woman dressed in a business suit) and unfamiliar (e.g., a woman dressed in a kimono) cultural items with either a home versus foreign or a regional vs. foreign accent contrast, but again, not a home versus regional contrast (Wagner et al., 2014). One explanation for these findings is that at least initially, children's representations are organized along a quantitatively defined continuum, anchored by their "home" accent, which forms the core set of their experiences, with regional accents treated as intermediate between the home and foreign accent (Wagner et al., 2014). With more experience, older children are able to reliably distinguish their own from another unfamiliar regional variety (Floccia et al., 2009) and make judgements in line with general social stereotypes. For example, children aged 9–10 years in Illinois and Tennessee, USA, judged northern-accented American English speakers to be more intelligent than those with southern-accented American English (Kinzler & DeJesus, 2013).

However, almost all of this work has focussed on monolingual children growing up in relatively homogenous communities. Yet one prediction of an exemplar-based approach is that with greater exposure to variation, and a more detailed bank of representations, more gradient, fine-grained representations will emerge. One possibility then is that children growing up in a multidialectal or multilingual environment like inner-city London, might be better able to extract and use variation in the speech signal, e.g., to group talkers, than those growing up in more homogenous environments. Few studies have investigated this question, and the results are mixed. For example, Jeffries (2016) found that monolingual English children aged 3–4 years, growing up in Yorkshire but with a parent from outside Yorkshire (i.e., who spoke with a different accent from that of the community), were better able to group talkers according to accent when the target variables contained different rather than the same phonemes (see also Floccia et al., 2009). Interestingly, children with parents from outside Yorkshire were also better able to recognize a familiar talker (in this case, a familiar teacher) using phonetic accent features. In contrast, Beck (2016) found that early exposure to multiple regional varieties, again through a parent with a different accent from that used in the local community, did not lead to better discrimination between different regional accents. One possibility is that this is because the task – an ABX discrimination task in which children heard and compared isolated words – was too easy and thus did not enable more fine-grained differences between the different listener groups to emerge. Indeed, differences in awareness did emerge, although perhaps not in the direction expected; children with parents from outside the community (Outsiders) performed more poorly at identifying whether or not an accent was local or non-local than their peers with parents born and raised in the community (Insiders), and being an Outsider negatively influenced the ability to recognize regional accent as a kind of linguistic variation between speakers. One possibility is that Outsiders could hear the variation in the input but, given their exposure to different varieties within the same place, did not equate speaking with a regional accent as being from a specific place (cf. Weatherhead et al., 2016). This lends further support to the notion that an understanding of social variation is derived from experience, and that how and when this emerges will vary between individuals as a result of early experience with language.

In this study we examine whether children raised monolingually or bilingually in a diverse community in London, differ in their ability to categorize talkers in three conditions: (1) home (London) versus unfamiliar regional accent (Yorkshire), (2) home versus unfamiliar foreign accent (Standard Singapore English, SSE), (3) unfamiliar regional versus unfamiliar foreign accent. Based on previous work, we hypothesize that though our monolinguals may outperform those in previous studies as a result of their increased exposure to accent variation (e.g., Wagner et al., 2014), they will likely find this task challenging. In contrast, we hypothesize that bilinguals will be better able to do this task, not just because they are exposed to more variation (regional and foreign-accented speech, as well as different languages) but also because this variation does meaningful work in their community, e.g., in signalling family relationships, and that this will mean that they are better able to categorize talkers even when the differences between them are relatively small (e.g., in the home vs. unfamiliar regional accent condition).

Method

Participants

Sixty children aged 5;4 to 7;11 years (mean 6;7 years) were recruited from North London primary schools in the neighbouring London boroughs of Brent and Barnet. None of the children who participated had any reported speech, hearing or language impairments, and prior to testing, all children obtained age-appropriate scores on the British Picture Vocabulary Scale 3 (Dunn & Dunn, 2009). Ethical approval for the study was granted by the UCL Ethics committee, and all children completed the study on an opt-in basis[1].

[1] Parents of all children in the target age group received a letter and information sheet informing them about the study and asking them if they would like their child to take part. Participation was on an opt-in basis: only those who returned the completed questionnaire and consent form to school and met the criteria took part.

Before testing, parents completed a language background questionnaire which provided basic details of their child's language background and exposure to other foreign languages or accents. Thirty were native monolingual English speakers (13 females, 17 males; 5;06 to 7;10 years, mean 6;05 years), and 30 were either bi- or multilingual (18 females, 12 males, 5;04 to 7;11 years, mean 6;08 years). Of these, only 5 were multilingual, i.e., they had acquired more than one additional language, and so, for ease, we will refer to this group as bilingual. Bilingual children were either simultaneous bilinguals (n = 13), who had acquired both English and any additional languages simultaneously from birth, or sequential bilinguals (n = 17) who had initially acquired their home language and had first received consistent exposure to English when attending nursery or pre-school. Bilingual children spoke a range of different languages (Gujarati, Urdu, Hindi, Polish, Romanian, Spanish, Portuguese), but none were bilingual in any of the other official languages spoken in Singapore (i.e., Mandarin, Malay, Tamil) or in any other Chinese language (e.g., Cantonese). The majority of children had been born and raised within the area of North London in which they were tested. A small number of bilingual children (n = 5) had either been born or spent time living outside the UK but had moved or returned to the UK to attend nursery school at the age of 3 years where they had lived ever since.

Information from the questionnaires was also used to inform the assignment of children to the different experimental conditions. For example, if a child fit the criteria for selection but his/her parents reported that he/she had had regular exposure to a variety of Northern English, the child was placed in the home-foreign condition. Regular exposure was defined as being through a close or near-relative (e.g., a parent, grandparent), or a caregiver with whom the child regularly interacted (e.g., a childminder). Only one child in the bilingual condition was reported to have regular exposure to Yorkshire English, and this child thus completed the home-foreign condition. This ensured that as far as possible, all children across all conditions were equally unfamiliar with both the regional and foreign accent.

Stimuli

Twelve speakers (2 male, 2 female per accent) recorded the stimuli for both the comprehension and categorization tasks. They were encouraged to read in an engaging style, as if reading a story to a child and all speakers were judged by a trained phonetician to have a representative accent. All speakers were recorded in a sound-attenuated booth at Chandler House, UCL, at a sampling rate of 44.1 kHz, 16-bit resolution, using a Rode NT-1A condenser microphone connected to a PC via an RME Fireface UC processor. Recordings were made in stereo and later converted to mono. Sentence stimuli were manually extracted and saved to individual .wav files with boundaries placed as close as possible to the onset and offset of each sentence. Stimuli were then band-pass filtered from 60 Hz to 20 kHz, equalized for intensity at 70 dB, and a 100-ms buffer added to the beginning and end of each file. All processing was completed in Praat (Boersma & Weenink, 2017).

Four monolingual native speakers from North London recorded the home accent (i.e., London) stimuli. The home accent was Popular London English (Wells, 1982), which is characterized by the use of glottal stops word-medially, TH-fronting, vocalized L, labialized R and differences in the realization of some vowels as compared with SSBE, e.g., the PRICE and MOUTH vowels. We chose this accent rather than SSBE because this has been shown to be highly intelligible, even for speakers of other regional accents (Pinet et al., 2015), and because although the children would have been familiar with SSBE, Popular London English is more representative of the accent used by them and those in their local community. Speakers were aged 24–49 years (mean 31 years) and had been born and raised in North London in the London boroughs of Camden and Islington. None had spent any significant amount of time living outside North London and were resident in the area at the time of recording. Although these speakers were from a different part of North London to the children who participated in the study, the accent is very similar to that used in the children's local communities, both of which border Camden.

The regional accent was Yorkshire English and was recorded by 4 monolingual native speakers from Yorkshire. Speakers were aged 22–50 years (mean 30 years). Of these, 3 speakers (1 male, 2 female) had been born and raised in West Yorkshire whilst the other speaker had been born and raised in North Yorkshire. All speakers were recorded in London. Three speakers had been living in London for 3 years, having moved there from Yorkshire to study at university, whilst the other had lived in London and the surrounding area for 15 years. Like other northern accents, Yorkshire English differs from southern English accents in production of BATH, which is produced with a short vowel, [æ], and STRUT, which is produced with either the FOOT vowel, [ʊ] (i.e., no STRUT-FOOT split) or a raised vowel similar to schwa, [ə]. Our speakers used both of these variants of the STRUT vowel, as is common in speakers of northern Englishes who have lived in the south of England (e.g., Evans & Iverson, 2007). Likewise, although our speakers also used some other features of Yorkshire English, e.g., a lowered /ɔː/ vowel, sometimes phonetically close to [ɒː] (Wells, 1982), they produced other vowels with more standard rather than regionally specific variants. For example, in these recordings they all produced FACE with a diphthong, [eɪ], rather than a monophthong, [eː] and GOAT, as [ɵʊ] (raised onset) or [əʊ], rather than as a monophthong [oː]. Thus, our speakers maintained key variants typical of their home community, which they used alongside some features of what Wells (1982) refers to as northern near-RP features.

The foreign accent was Singaporean English and was recorded by 4 native speakers of Singaporean English aged 22–26 years (mean 23 years). All speakers had been born and raised in Singapore and had been living in London for 4 years at the time of the recordings, where they had moved for university. All speakers had acquired Mandarin initially but had been exposed to English from an early age and had completed their education in Singapore bilingually in English and Mandarin. Key features include what has traditionally been referred to as syllable-timed rhythm (Trudgill & Hannah, 2002), avoidance of schwa in unstressed syllables (Wells, 1982), neutralization of vowel length distinctions (Deterding, 2005), TH-stopping (Bao, 1998) and consonant cluster reduction (Lim, 2004). Our speakers displayed all of these features and also differed from native English speakers in their use of connected speech features, such as linking and intrusive /r/, which are absent from SSE (Trudgill & Hannah, 2002), and realization of word-final consonants, which in SSE are typically voiceless, leading to minimal pairs like "niece" and "knees" in SSBE becoming homophones in SSE (Lim, 2004).

Stimuli were selected to include key phonetic differences between the different accents but were not intended to be highly confusable. Differences between London and SSE affected many different phonetic features, but all sentences were also designed to include key differences between London and Yorkshire English, i.e., a BATH or STRUT vowel (see the Appendix for a full set of target and distractors). The sentence stimuli for the comprehension task consisted of six 1- to 2-sentence phrases, four of which were themed to include characters from *The Gruffalo* (Donaldson & Scheffler, 1999). The stimuli for the categorization task consisted of twelve 1- to 2-sentence passages. Of these, 6 were taken from the popular children's book *Mrs Plug the Plumber* (Ahlberg & Wright, 1980) and 6 were original. Vocabulary was selected to be age-appropriate; average acquisition of each word was either at or below the age of 5 years as detailed in the Age of Acquisition database (Kuperman et al., 2012).

Procedure

All testing took place in a quiet room in the child's school and was carried out by the second author, who had grown up in North London and spoke with a North London accent. During the experiment, the child and experimenter were seated next to each other at a table. Stimuli were presented at a comfortable listening level via a laptop PC over headphones, worn by both the experimenter and child, using the Experiment MFC interface in Praat (Boersma & Weenink, 2017). The child gave his/her responses using picture response cards, and these were simultaneously logged by the experimenter via the Experiment MFC interface (Boersma & Weenink, 2017). The child was seated so he/she was unable to see the computer screen. Presentation of stimuli was pseudo-randomized across participants and conditions. Children completed the comprehension and then the categorization task in a single testing session lasting 10–15 min.

Comprehension Task

The task was a 4-alternative forced choice task, in which children selected the picture which they thought best matched the sentence they had heard. Previous work has shown that adults are able to quickly tune in to an unfamiliar accent and so the purpose of this task was both to famil-

Fig. 1. Pictures used in trial 4 of the comprehension task. The target picture shows a ball in the grass, with the other 3 pictures distractors.

iarize children with the novel accent as well as to test that they were able to understand it, i.e., that any differences in performance in the categorization task were not because they could not understand the talker.

Children completed 6 trials in the non-home dialect, i.e., either the regional (Yorkshire) or foreign (SSE) accent, produced by 2 talkers (1 male, 1 female). Before beginning the task, children were given a small booklet which contained separate response sets for each of the 6 trials. Each set consisted of 4 picture cards (1 target, 3 distractors) which were attached by Velcro, with only one set visible during a given trial (Fig. 1). The position of the pictures in each set was randomized across participants to reduce any potential effects of side bias on picture selection. The experimenter explained to the child that he/she would hear someone speaking and that it was his/her job to listen carefully, select the picture that the speaker had asked for or described (i.e., remove it from the booklet) and give the picture card to the experimenter. For example, on hearing "Can you see the ball in the grass?" (trial 4), children saw pictures of a doll, a teddy, a bowl and a ball in the grass and had to choose the picture with the ball (Fig. 1). To keep the child "on task," the first, third and fifth trials were preceded by appropriate attention-grabbing audio files. For example, trial 5 ("His favourite food is a glass of mouse milkshake"), was preceded by the sound of liquid being sucked through a straw. All clips were downloaded from the Internet and converted to mono .wav files in Praat (Boersma & Weenink, 2017). A trial was considered complete when a picture had been selected and given to the experimenter. Children were praised uniformly throughout.

Categorization Task

The task was divided into 2 parts; training in group membership (4 trials) and a test phase in which children completed 6 categorization trials, interspersed with 2 reminder training trials which followed the 2nd categorization trial.

Children were first introduced to two soft toys, the Mouse and Gruffalo's Child. Both are well-known characters from the children's picture book *The Gruffalo's Child* (Donaldson & Scheffler, 2001; also an animated short film) and were familiar to all the children. Each toy represented one of the two accents (i.e., home and foreign, home and regional, regional and foreign) with the assignment of character to accent counterbalanced across participants: both soft toys were on view during the whole of the categorization task. Children were told that the characters' families had got lost in the deep, dark wood (the setting for the story) and that their job was to help each of them find the correct family (i.e., mouse or Gruffalo's child). However, because the wood was so dark they could not see them, and so they would have to listen carefully to their voices instead. The children were then introduced to how the mouse and Gruffalo's child families talked. At the be-

DOI: 10.1159/000493983

Table 1. Mean proportion correct (standard deviation in parentheses) for the regional and foreign accents in the comprehension task, split by language background (monolingual, bilingual)

	Regional	Foreign
Monolingual	0.91 (0.09)	0.91 (0.14)
Bilingual	0.88 (0.13)	0.86 (0.15)

ginning of each training trial, the experimenter held up the appropriate soft toy, and said "This is what the mouse/Gruffalo's child family sounds like!". A sound file was then played whilst the experimenter held up and jiggled the appropriate soft toy. Children were trained in this way with 2 sentences (1 male, 1 female) from each accent (i.e., 4 different talkers) before continuing to the test phase, in which they completed a 2-alternative forced choice task where they responded whether the talker they heard belonged to the Gruffalo's child or mouse family. Before beginning the test phase, children were reminded that they would hear a sentence from a character lost in the deep, dark wood, and that their job was to say which family the speaker belonged to by selecting the appropriate picture (mouse or Gruffalo's child) from the response booklet. There were 6 test trials (3 per accent). For each accent (i.e., 3 trials), the speakers were 1 talker who was familiar from the training and 2 novel talkers. The familiar talker was also used in the reminder trials. As in the comprehension task, a test trial was considered complete when a picture had been selected and given to the experimenter. Children were praised uniformly throughout.

Results

Comprehension

Table 1 shows performance (proportion correct) for each accent in the comprehension task (regional, foreign) for monolingual and bilingual children. All children performed very well at this task, and there did not appear to be any large differences in overall performance between monolingual and bilingual children, though monolinguals had a higher mean comprehension score than bilinguals.

These observations were confirmed in a series of analyses. First, a series of 1-sample t tests confirmed that all children performed above chance in all conditions: regional accent, monolinguals – $t(14) = 25.58$, $p < 0.001$, bilinguals – $t(14) = 21.75$, $p < 0.001$; foreign accent, monolinguals – $t(14) = 40.12$, $p < 0.001$, bilinguals – $t(14) = 25.66$, $p < 0.001$. Secondly, a univariate ANOVA with accent (regional, foreign) and language background (monolingual, bilingual) coded as independent variables, and proportion correct as the dependent variable, demonstrated that there were no main effects or interaction of accent or language background, $p > 0.05$, confirming that monolinguals and bilinguals performed similarly with the two accents.

Previous work has shown that children's identification of sentences presented in quiet in a foreign accent is significantly poorer than in a home accent (Bent & Atagi, 2015), and so it might seem somewhat surprising that our children performed so well in this task. However, our task was designed to be relatively easy to enable children to tune into the accents and to build confidence in doing the more difficult categorization task. It is thus possible that children did experience some difficulties in processing the unfamiliar accents, but that the use of pictures may have enabled them to make use of context in selecting the correct answer (cf. Bent, 2014).

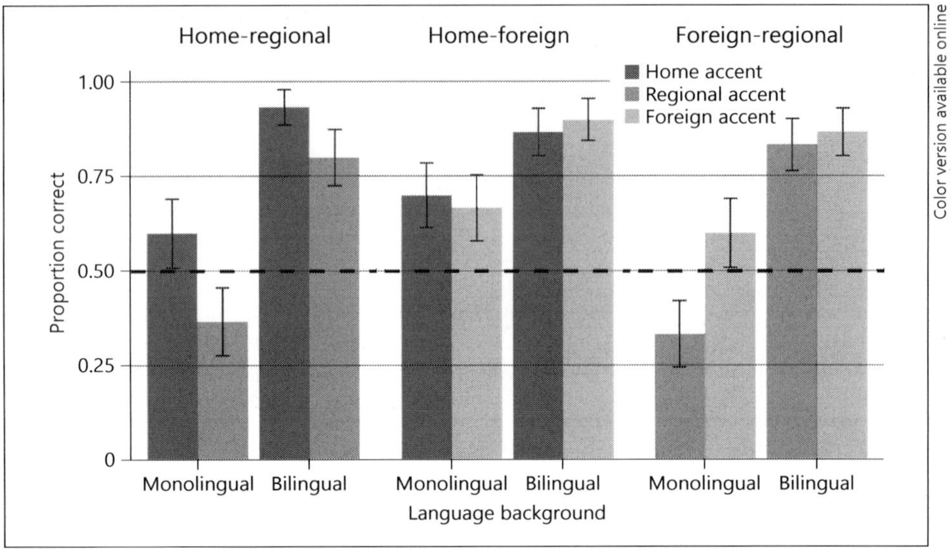

Fig. 2. Mean categorization score (proportion correct) for monolingual and bilingual children in each condition (home-regional, home-foreign, regional-foreign) split by accent. Note that each bar represents the average score across 3 trials.

Categorization Task

Overall Performance

Figure 2 displays the mean proportion correct for each accent (home, regional, foreign) in each condition (home-regional, home-foreign, regional-foreign) split by language background (monolingual, bilingual). Responses were coded as correct when children selected the soft toy that matched that accent. That is, if children were trained that the mouse's family spoke with the home (Popular London) accent and the Gruffalo spoke with the foreign (SSE) accent, answering that a speaker who speaks with a London accent belonged to the mouse's family was coded as a correct response, whilst answering that he/she belongs to the Gruffalo's family (foreign accent) was coded as an incorrect response. Inspection of the data suggests that bilinguals do better than monolinguals overall, performing at near ceiling level in all conditions. In contrast, monolinguals appear to find the task more difficult; although they do better in the home-foreign condition than in the home-regional and regional-foreign conditions, they perform more poorly than bilinguals overall.

To assess children's overall performance, an overall categorization score (i.e., the mean categorization score across both dialects in the pair) was calculated for each child in each group (monolingual, bilingual) for each condition (home-regional, home-foreign, regional-foreign) and compared to chance. Bilinguals performed above chance in all conditions (Bonferroni-corrected for multiple comparisons): home-regional $t(9) = 10.94$, $p < 0.001$, home-foreign $t(9) = 10.73$, $p < 0.001$, regional-foreign $t(9) = 9.04$, $p < 0.001$. In contrast, monolinguals performed above chance in the home-foreign condition [$t(9) = 4.01$ $p < 0.01$], but not in the home-regional and regional-foreign conditions, $p > 0.05$.

Table 2. Summary of the results of the regression model for accent categorization

	β	SE	z	p
Intercept	0.733	0.297	2.463	0.014
Condition (F-R)	–0.840	0.410	–2.047	0.041
Condition (H-R)	–0.782	0.403	–1.942	0.052
Language (b)	1.262	0.490	2.575	0.010
Condition (F-R): language (b)	0.621	0.663	0.937	0.349
Condition (H-R): language (b)	0.696	0.673	1.033	0.301

F-R, foreign-regional; H-R, home-regional; b, bilingual. Baselines for predictor variables: home-foreign for condition; monolingual for language background. Figures represent estimates (β), standard errors (SE), Wald statistics (*z*-values) and *p* values.

A mixed effect logistic regression model was built with the binomial response (correct/incorrect) as the dependent variable, condition (home-foreign, foreign-regional, home-foreign) and language background (monolingual, bilingual) as fixed factors and "participant" and "speaker" as crossed random effects. Treatment coding was used for both factors, with "home-foreign" as the reference level for condition and "monolingual" as the reference level for language background. For monolingual listeners, the model revealed a significant contrast between the home-foreign ($M_{prop} = 0.683$) and regional-foreign ($M_{prop} = 0.467$) conditions and a marginal significant contrast between the home-foreign ($M_{prop} = 0.683$) and home-regional ($M_{prop} = 0.483$) conditions (Table 2). There was a significant effect of language background, which indicated that bilinguals performed better than monolinguals (Fig. 1). The interaction between condition and language background was not significant. To investigate whether bilinguals differed from monolinguals in each condition, pairwise post hoc tests were calculated using the lsmeans package (Lenth, 2016) in R (R Core Team, 2018), adjusting for multiple comparisons using the Tukey method. The tests indicated that bilinguals outperformed monolinguals in all conditions (home-foreign: $\beta = -1.262$, SE = 0.490, $z = -2.575$, $p = 0.01$; foreign-regional: $\beta = -1.883$, SE = 0.447, $z = -4.208$, $p < 0.001$; home-regional: $\beta = -1.958$, SE = 0.463, $z = -4.225$, $p < 0.001$).

As noted by Girard et al., (2008; see also Floccia et al., 2009), measures of correct responses do not exclude potential effects of response bias, that is, the possibility that children might over- or underestimate the number of unfamiliar/untrained accented sentences in the categorization task. To assess whether or not children were able to detect the unfamiliar/untrained accent using a bias-free measure, separate measures of sensitivity (A') and bias (B''_D) were computed using signal detection analysis (see Girard et al., 2008; Floccia et al., 2009). Sensitivity (A') was used rather than d' (Grier, 1971) because it does not rely on the strict assumptions of normality and equal variance of signal and noise distributions. Additionally, unlike d', this measure, which varies between 0 and 1, can be computed for extreme values. Thus, a score of 1.0 represents a 100% correct hit-rate (i.e., no false alarms), and a value of 0.5 indicates chance performance. In this case, a hit was defined as correct categorization of the home accent in the home-foreign and home-regional conditions, and the regional

Table 3. Mean measures of sensitivity (A') and bias (B"$_D$) for monolingual and bilingual children calculated for each condition (home-regional, home-foreign, regional-foreign)

	Home-regional		Home-foreign		Regional-foreign	
	A'	*B"*$_D$	*A'*	*B"*$_D$	*A'*	*B"*$_D$
Monolingual	0.46	−0.32	0.76	−0.08	0.42	0.36
Bilingual	0.92	−0.42	0.93	−0.08	0.91	0.2

accent in the regional-foreign condition. A false alarm occurred when the regional (home-regional condition) or foreign accent (home-foreign and regional-foreign conditions) was mistakenly identified as the home or regional accent. Response bias, measured as B''_D, is independent from A' and ranges from −1 to 1. Negative values indicate a liberal response bias, and positive values indicate a conservative bias (Donaldson, 1992).

Table 3 displays the average sensitivity and bias for monolingual and bilingual children in each condition (home-regional, home-foreign, regional-foreign). The measure of sensitivity gives a similar pattern of results to the overall categorization score; bilingual children are highly sensitive in all conditions, but monolinguals have only above-chance levels of sensitivity in the home-foreign condition. Potential differences in performance according to language background were investigated in a univariate ANOVA with language background and condition coded as between-subject variables. There was a significant main effect of language background, $F(1, 49) = 62.34$, $p < 0.001$, confirming that bilinguals showed greater sensitivity to accent differences than monolinguals. Additionally, there was a main effect of condition, $F(2, 49) = 7.01$, $p < 0.001$, and interaction of language background and condition, $F(2, 49) = 5.47$, $p < 0.01$. Again, this reflects the finding that monolinguals had greater sensitivity in the home-foreign condition than in the home-regional or regional-foreign conditions, with bilinguals having similarly high sensitivity across all 3 conditions.

In terms of bias, monolingual and bilingual children showed similar tendencies in their responses in the different conditions (Table 3). All children had a negative bias in the home-regional condition, indicating that they had a tendency to answer "home" (i.e., to select the soft toy associated with the London accent), whilst in the regional-foreign condition they had positive bias, indicating a tendency to answer "foreign" (i.e., to select the soft toy associated with SSE). Given the success with which children were able to complete the task in the home-foreign condition, it is not surprising that neither monolingual nor bilingual children showed evidence of bias in this condition. These observations were confirmed in a univariate ANOVA on bias, with language background and condition coded as between-subject variables. There was no significant main effect of language background or interaction of language background and condition. However, there was a significant effect of condition, $F(2, 44) = 3.27$, $p < 0.05$, confirming that responses, regardless of language background, showed differing effects of bias in the 3 conditions. Thus, in the home-regional condition, all children, regardless of language background, had a bias towards answering that the talker had the "home" accent, whilst in the regional-foreign condition, all children had a bias towards answering that the talker had the "foreign" accent.

Discussion

Previous research with monolingual children has shown that the ability to explicitly categorize talkers according to regional accent begins to emerge at around 5 years, and that initially they can only successfully categorize speakers when the acoustic-phonetic differences between the accents are maximized (Floccia et al., 2009; Girard et al., 2008; Wagner et al., 2014). This led researchers to hypothesize that children have gradient representations of accent variation, with their home accent, the one that is experienced the most, forming the core set of experiences and other varieties identified relative to this (Wagner et al., 2014). Children are thus thought to perform poorly in categorizing talkers from their home versus an unfamiliar regional accent community because even though these accents might be phonetically quite different, they are interpreted by the children as being similar enough that they cannot be well differentiated. In contrast, foreign accents are hypothesized to go well beyond what children will accept as similar, and so they are able to categorize these as separate from their own, native accent (Wagner et al., 2014).

Based on these findings, our initial hypothesis was that with greater experience of variation, children would develop a more detailed, gradient representation of variation within their community, such that our children, growing up in a community where they are frequently exposed to regional and foreign-accented speech, would perform well on an accent categorization task. Specifically, we hypothesized that our monolingual children would perform better than those in previous studies, who were growing up in more homogenous communities (Girard et al., 2008; Wagner et al., 2014), and that our bilingual children, who are exposed to the greatest amount of variability, would perform still more successfully than both of these groups. Our results supported our second prediction but not the first. Bilingual children performed best and were able to categorize talkers successfully in all 3 conditions (home-regional, home-foreign, regional-foreign). In contrast, although all children were able to understand the different accents, our monolingual children were only able to categorize talkers with above-chance accuracy in the home-foreign condition, replicating findings from these previous studies but with a different population.

Why did our monolingual children perform more poorly on the categorization task than our bilinguals? One interpretation is that these results simply lend support for the gradient representation hypothesis outlined above. One could imagine that at this stage of development, children may be more sensitive to the broader and less predictable differences that typically occur in foreign-accented speech (Girard et al., 2008) and that for our monolingual children, with potentially less exposure to foreign-accented varieties than our bilingual children, these differences were particularly noticeable. However, the foreign accent in our study was SSE, an established World English variety, with systematic patterns of variation which were well exemplified by our talkers, who had acquired this variety in childhood (cf. Wagner et al., 2014). All our speakers were highly fluent, and it is therefore highly unlikely that they differed from speakers of our other varieties in their consistency of phonological patterning.

Another possibility is that success at this task depends on experience with different accents. For example, in exemplar-theoretic models, adults have been hypothesized to be able to identify talkers according to their regional accent by mapping to stored memories of similar talkers (e.g., Johnson, 1997). The regional and foreign accents used in this study were selected to be unfamiliar to the children, and it is possible that our monolingual children were unable to successfully categorize talkers according to accent as a result of their inexperience with these particular accents. That is not to say that they heard the regional and home accent as equivalent – our monolinguals performed poorly in the regional versus foreign accent condition but not the home versus foreign accent condition – but rather that the regional accent was within the "noise" tolerance for the home accent and that their sparse representations for the regional and foreign accent meant that they were unable to separate them reliably (Wagner et al., 2014). One obvious problem with this explanation is that our bilinguals were selected to be equally unfamiliar with the regional and foreign accents used here; although they likely had more experience of variation overall, they did not have experience with these particular accents. Likewise, although we did not conduct any tests of cognitive abilities, there were no significant differences in vocabulary in monolingual and bilingual children, with all achieving age-appropriate scores on the British Picture Vocabulary Scale. It therefore seems unlikely that our monolingual children were unable to do this task either because it is too difficult for children at this age (cf. Jones et al., 2017) or because they had no experience with those accents.

Perhaps a better explanation is that successful categorization of talkers according to regional accent involves not only development of the ability to track acoustic-phonetic differences between talkers, consistent with the gradient representation hypothesis, but also an understanding of how patterns of variation are used meaningfully within a community. In turn, this enables children to associate patterns of variation with a given attribute. In support of this, previous research has shown that bilinguals have an enhanced performance on phonological tasks in comparison to their monolingual counterparts (Bialystok et al., 2003; Campbell & Sais, 1995; Yelland et al., 1993), enhanced attentional processing (Kovacs & Mehler, 2009) and greater linguistic flexibility (Bialystok, 1986). Bilinguals arguably also have a greater need to use linguistic information in order to navigate relationships within their community (see McCarthy et al., 2014). Increased exposure to variation in a community where such variation serves to differentiate talkers as belonging to different sections of that community may thus promote sensitivity to talker variation, such that children growing up bilingually in a diverse, multilingual community develop the ability to extract, store and use talker variation in speech processing in a more fine-grained way, earlier in development.

This explanation is in part consistent with Labov's (1964) proposal that there are multiple stages in the development of the processing of indexical variation. In this model, children are initially thought to focus on learning the basic grammar of their language from their parents/caregivers, with the ability to understand the social significance of dialect variation only emerging in early adolescence (cf. Jones et al., 2017). Whilst our results support the idea that children's representations are likely initially influenced primarily by their core set of experiences with their home dialect(s), they also suggest that social context plays an important role in the development of sociolinguistic awareness. The fact that our bilinguals, exposed to greater amounts of variation than our monolinguals, were highly successful at this task indicates that although all children likely have the capacity to perceive differences between talkers early in development, the ability to do this might present differently and at a different stage in development, depending on the child's linguistic environment. Crucially, our evidence suggests that for bilinguals growing up in a diverse environment, an understanding of the social significance of dialect variation may begin to emerge earlier than previously thought. Of course, not all bilingual children grow up in this kind of environment. Others grow up in communities where they are exposed to a single, dominant variety of the host country language in their community and, though they speak their other language at home with a caregiver and other close family members, may only have limited experience of it outside this setting (cf. Grosjean, 1998). This differs from the experience of the bilingual children growing up in dense immigrant communities like the one tested here, where children have increased exposure not only to different native-speaker varieties of English (e.g., through their monolingual peers, teachers) but also to foreign-accented varieties (McCarthy et al., 2013). In this case, we would predict that those bilinguals, exposed to less variation, would perform more like our monolinguals, with the ability to categorize talkers according to spoken accent emerging later in development.

What is being encoded? Previous research has suggested there is a developmental progression from specific and concrete knowledge (phonetically rich, potentially talker-specific) to more abstract knowledge of sound structure (generalizations), with age and exposure playing an important role in the development of an understanding of the social significance of variation (cf. Munson, 2010, Beckman et al., 2007). Our results provide some tentative support for this view; despite being unfamiliar with the accents tested, our bilingual children were able to successfully categorize talkers, perhaps because they had developed representations in which particular speaker attributes were associated with more fully developed macrosociological categories. However, the exact nature of these representations requires further exploration, and whilst our results highlight the need for an important role for social context, we would not want to exclude a role for a phonetic-phonological explanation. For example, one could imagine that If listeners' representations develop to be tolerant of variation, at least to some degree, then as well as facilitating perceptual adaptation by making it easier to map phonetic variation to underlying phonological (i.e., abstract) categories (Shaw et al., 2018), they might also facilitate mapping of variation across individual talkers to macrosociological constructs. In developmental terms, this might mean that as children shift from phonetic to phonological processing (Nathan et al., 1998), they may be better able to spot and extract patterns of variation, and map these to social categories based on their developing understanding of their social world.

Conclusion

This study has shown that early experience with speech affects children's ability to use variation in an explicit categorization task. Despite the fact that children were exposed to a small number of talkers and heard only a small number of sentences, bilingual children were able to categorize talkers in all 3 accent pairings. In contrast, monolingual children were only able to categorize talkers in the home-foreign conditions, where the differences between the accents were maximized. Although these findings are consistent with an exemplar-based account, they could also be interpreted in the context of a gradient representation hypothesis, with the home accent forming the core set of experiences and other accents identified relative to this. They further suggest an important role for social context such that the development of the ability to extract and use patterns of variation to categorize talkers may differ according to a child's linguistic environment. Knowledge of phonetic and social variation is thus likely emergent, with any abstract categories formed in part at least, from social and phonetic knowledge gained and filtered through experience. Thus, just as phonetic perception becomes specialized for one's native language(s) early in life (Kuhl et al., 2006; Werker & Tees, 1984), so too might sociolinguistic awareness, with perception filtered through the lens of an individual's understanding of their social world.

Appendix

Stimuli used in the comprehension and categorization tasks, with response options given where relevant.

i Comprehension task

	Stimulus sentence	Response options (pictures)
Trial		
1	One of his favourite foods is a buttered mouse roll	Mouse in a roll, fox in a roll, owl in a roll, snake in a roll
2	The Gruffalo needs a bath, show me the soap	Bath, soap, rubber duck and sponge
3	One of his favourite foods is a jammy doughnut	Bubble gum, doughnut, fairy cake, chocolate bar
4	Can you see the ball in the grass?	Ball in grass, patch of grass, teddy in grass, flower growing in the grass
5	His favourite drink is a glass of mouse milkshake	Glass of milkshake, glass of orange juice, glass of fizzy pop, carton of juice
6	Show me the giraffe with the purple scarf	Giraffe with red scarf, green scarf, purple scarf, blue scarf

ii Categorization task

	Sentence
Training trial	
1	Oh no! Mummy dropped her money in the puddle! And her gloves are soaking wet!
2	The children ran through the grass and jumped in the puddle and they got their coats very muddy
3	One night there was a storm. The ship hit a rock and a hole appeared in its side
4	The next night there was a big celebration party. Mrs. and Mr. Plug danced on deck under the stars
Reminder trials	
1	The lady gave her a cup of tea. When she had finished the lady gave her some money, Mrs. Plug set off home
2	Mrs. Plug got on with the job mending the hole in the boat. Mr. Plug helped her. They both got soaking wet

	Stimuli
Test trial	
1	Mrs. Plug went home. She told Mr. Plug about the robber and the rich man. She showed him the reward
2	Mrs. Plug had a useful bag. It had a saw, a hammer, a spanner, a purse, a comb, a lipstick, a plunger, a blow torch and a few other things in it
3	One night there was some trouble in a lady's bathroom. A plumber was needed. The neighbours said send for Mrs. Plug!
4	The puppy ate all of the boy's lunch without leaving a crumb. Now his tummy was very, very full
5	The monkey stuck out his tongue and licked the glass and all of the children laughed
6	The boy put on his woolliest gloves and rolled a big snowball. He threw the snowball and it hit the window

Acknowledgements

This paper presents data that were collected by Amy Madigan as part of her undergraduate dissertation at University College London in partial fulfilment of the BSc Speech Communication. The writing of this paper has benefitted from discussions with many people, in particular Paul Foulkes, Ella Jeffries and Josiane Riverin-Coutlée, and participants of the symposium on "Children's understanding and use of sociolinguistic variation in language acquisition" at the 14th International Congress for the Study of Child Language, organized by Laura Wagner and Cynthia Clopper, where aspects of this work were presented. We would also like to thank all the parents and children and their schools who generously gave of their time to take part in this research.

Disclosure Statement

The authors have no conflicts of interest to declare.

References

Adank, P., Evans, B. G., Stuart-Smith, J., & Scott, S. K. (2009). Comprehension of familiar and unfamiliar native accents under adverse listening conditions. *Journal of Experimental Psychology. Human Perception and Performance, 35*(2), 520–529. doi:10.1037/a0013552

Ahlberg, A., & Wright, J. (1980). *Mrs Plug the Plumber.* London, England: Puffin.

Bao, Z. (1998). The sounds of Singapore English. In J. A. Foley, T. Kandiah, Z. Bao, A. F. Gupta, L. Alsagoff, C. L. Ho, . . . Bokhorst-Heng, W. (Eds.), *English in New Cultural Contexts: Reflections from Singapore* (pp. 127–151). Singapore: Oxford University Press.

Bayard, D., Gallois, C., Weatherall, A., & Pittam, J. (2001). Pax Americana? Accent attitudinal evaluations in New Zealand, Australia and America. *Journal of Sociolinguistics, 5*(1), 22–49. doi:10.1111/1467-9481.00136

Beck, E. (2016). What It Means to Be an Outsider: How Exposure to Regional Variation Shapes Children's Awareness of Regional Accents in Their Native Language. In A. Babel (Ed.), *Awareness and Control in Sociolinguistic Research* (pp. 104–122). Cambridge: Cambridge University Press. doi:10.1017/CBO9781139680448.007

Beckman, M. E., Munson, B., & Edwards, J. (2007). Vocabulary growth and developmental expansion of types of phonological knowledge. In J. Cole & J. I. Hualde (Eds.), *Laboratory Phonology 9* (pp. 241–264). New York: Mouton de Gruyter.

Bent, T. (2014). Children's perception of foreign-accented words. *Journal of Child Language, 41*(6), 1334–1355. doi:10.1017/S0305000913000457

Bent, T., & Atagi, E. (2015). Children's perception of nonnative-accented sentences in noise and quiet. *The Journal of the Acoustical Society of America, 138*(6), 3985–3993. doi:10.1121/1.4938228
Best, C. T., Tyler, M. D., Gooding, T. N., Orlando, C. B., & Quann, C. A. (2009). Development of phonological constancy: Toddlers' perception of native- and Jamaican-accented words. *Psychological Science, 20*(5), 539–542. doi:10.1111/j.1467-9280.2009.02327.x
Bialystok, E. (1986). Factors in the growth of linguistic awareness. *Child Development, 57*(2), 498–510. doi: 10.2307/1130604
Bialystok, E., Majumder, S., & Martin, M. M. (2003). Developing phonological awareness: Is there a bilingual advantage? *Applied Psycholinguistics, 24*(01), 27–44. doi:10.1017/S014271640300002X
Boersma P, Weenink D (2017): *Praat: Doing phonetics by computer.* Retrieved from: www.praat/org
Campbell, R., & Sais, E. (1995). Accelerated metalinguistic (phonological) awareness in bilingual children. *British Journal of Developmental Psychology, 13*(1), 61–68. doi:10.1111/j.2044-835X.1995.tb00664.x
Campbell-Kibler, K. C. (2016). Towards a Cognitively Realistic Model of Meaningful Sociolinguistic Variation. In A. Babel (Ed.), *Awareness and Control in Sociolinguistic Research* (pp. 123–151). Cambridge: Cambridge University Press. doi:10.1017/CBO9781139680448.008
Cheshire, J., Kerswill, P., Fox, S., & Torgersen, E. (2011). Contact, the feature pool and the speech community: The emergence of Multicultural London English. *Journal of Sociolinguistics, 15*(2), 151–196. doi:10.1111/j.1467-9841.2011.00478.x
Clarke, C. M., & Garrett, M. F. (2004). Rapid adaptation to foreign-accented English. *The Journal of the Acoustical Society of America, 116*(6), 3647–3658. doi:10.1121/1.1815131
Clopper, C. G., & Bradlow, A. R. (2008). Perception of dialect variation in noise: Intelligibility and classification. *Language and Speech, 51*(Pt 3), 175–198. doi:10.1177/0023830908098539
Cristia, A., Seidl, A., Vaughn, C., Schmale, R., Bradlow, A., & Floccia, C. (2012). Linguistic processing of accented speech across the lifespan. *Frontiers in Psychology, 3*, 479. doi:10.3389/fpsyg.2012.00479
Darcy, I., & Krüger, F. (2012). Vowel production and perception in Turkish children acquiring L2 German. *Journal of Phonetics, 40*(4), 568–581. doi.org/10.1016/j.wocn.2012.05.001
Deterding, D. (2005). Emergent patterns in the vowels of Singapore English. *English World-Wide, 26*(2), 179–197. doi:10.1075/eww.26.2.04det
Docherty, G. J., & Foulkes, P. (2014). An evaluation of usage-based approaches to the modelling of sociophonetic variability. *Lingua, 142*, 42–56.
Donaldson, J., & Scheffler, A. (2001). *The Gruffalo's child.* London, England: Macmillan Children's Books.
Donaldson, W. (1992). Measuring recognition memory. *Journal of Experimental Psychology. General, 121*(3), 275–277. doi:10.1037/0096-3445.121.3.275
Dunn, D. M., & Dunn, L. M. (2009). *The British Picture Vocabulary Scale. National Foundation for Educational Research in England and Wales.; GL Assessment (Firm)* (3rd ed.). London: GL Assessment.
Evans, B. G., & Iverson, P. (2007). Plasticity in vowel perception and production: A study of accent change in young adults. *The Journal of the Acoustical Society of America, 121*(6), 3814–3826. doi: 10.1121/1.2722209
Floccia, C., Butler, J., Girard, F., & Goslin, J. (2009). Categorization of regional and foreign accent in 5- to 7-year-old British children. *International Journal of Behavioral Development, 33*(4), 366–375. doi: 10.1177/0165025409103871
Floccia, C., Goslin, J., Girard, F., & Konopczynski, G. (2006). Does a regional accent perturb speech processing? *Journal of Experimental Psychology. Human Perception and Performance, 32*(5), 1276–1293. doi:10.1037/0096-1523.32.5.1276
Foulkes, P., & Docherty, G. J. (2006). The social life of phonetics and phonology. *Journal of Phonetics, 34*(4), 409–438. doi:10.1016/j.wocn.2005.08.002
Girard, F., Floccia, C., & Goslin, J. (2008). Perception and awareness of accents in young children. *British Journal of Developmental Psychology, 26*(3), 409–433. doi:10.1348/026151007X251712
Goldinger, S. (1997). Words and voices: Perception and production in an episodic lexicon. In K. Johnson & J. Mullennix (Eds.), *Talker variability in speech processing* (pp. 33–66). San Diego: Academic Press.
Grier, J. B. (1971). Nonparametric indexes for sensitivity and bias: Computing formulas. *Psychological Bulletin, 75*(6), 424–429. doi:10.1037/h0031246
Grosjean, F. (1998). Studying bilinguals: Methodological and conceptual issues. *Bilingualism: Language and Cognition, 1*(2), 131–149. doi:10.1017/S136672899800025X
Hawkins, S. (2003). Roles and representations of systematic fine phonetic detail in speech understanding. *Journal of Phonetics, 31*(3-4), 373–405. doi:10.1016/j.wocn.2003.09.006
Hawkins, S., & Smith, R. (2001). Polysp: A polysystemic, phonetically-rich approach to speech understanding. *Rivista di Linguistica, 13*, 99–188.

Hay, J., & Drager, K. (2010). Stuffed toys and speech perception. *Linguistics, 48*(4), 865–892. doi:10.1515/ling.2010.027

Iverson, P., & Evans, B. G. (2009). Learning English vowels with different first-language vowel systems II: Auditory training for native Spanish and German speakers. *The Journal of the Acoustical Society of America, 126*(2), 866–877. doi:10.1121/1.3148196

Jeffries E (2016). *Children's developing awareness of regional accents: A socioperceptual investigation of pre-school and primary school children in York* (Unpublished PhD thesis), University of York, UK.

Johnson, K. (1997). Speech perception without speaker normalization: An exemplar model. In K. Johnson & J. Mullennix (Eds.), *Talker variability in speech processing* (pp. 145–165). San Diego: Academic Press.

Jones, Z., Yang, Q., Wagner, L., & Clopper, C. G. (2017). The development of dialect classification across the lifespan. *Journal of Phonetics, 60*, 20–37. doi:10.1016/j.wocn.2016.11.001

Kinzler, K. D., & DeJesus, J. M. (2013). Northern = smart and Southern = nice: The development of accent attitudes in the United States. *Quarterly Journal of Experimental Psychology, 66*(6), 1146–1158. doi: 10.1080/17470218.2012.731695

Kovács, Á. M., & Mehler, J. (2009). Cognitive gains in 7-month-old bilingual infants. *Proceedings of the National Academy of Sciences of the United States of America, 106*(16), 6556–6560. doi:10.1073/pnas.0811323106

Kuhl, P. K., Stevens, E., Hayashi, A., Deguchi, T., Kiritani, S., & Iverson, P. (2006). Infants show a facilitation effect for native language phonetic perception between 6 and 12 months. *Developmental Science, 9*(2), F13–F21. doi:10.1111/j.1467-7687.2006.00468.x

Kuperman, V., Stadthagen-Gonzalez, H., & Brysbaert, M. (2012). Age-of-acquisition ratings for 30,000 English words. *Behavior Research Methods, 44*(4), 978–990. doi:10.3758/s13428-012-0210-4

Labov, W. (1964). Stages in the acquisition of standard English. In R. Shuy (Ed.), *Social dialects and language learning* (pp. 77–104). Champaign, IL: National Council of Teachers of English.

Lenth, R. V. (2016). Least-squares means: The R package lsmeans. *Journal of Statistical Software, 69*(1), 1–33. doi:10.18637/jss.v069.i01

Lim, L. (2004). *Singapore English: A grammatical description (varieties of English around the world G33).* Amsterdam, Philadelphia: John Benjamins. doi:10.1075/veaw.g33

Maye, J., Aslin, R. N., & Tanenhaus, M. K. (2008). The weckud wetch of the wast: Lexical adaptation to a novel accent. *Cognitive Science, 32*(3), 543–562. doi:10.1080/03640210802035357

Mayr, R., Howells, G., & Lewis, R. (2015). Asymmetries in phonological development: The case of word-final cluster acquisition in Welsh-English bilingual children. *Journal of Child Language, 42*(1), 146–179. doi:10.1017/S0305000913000603

McCarthy, K. M., Evans, B. G., & Mahon, M. (2013). Acquiring a second language in an immigrant community: The production of Sylheti and English stops and vowels by London-Bengali speakers. *Journal of Phonetics, 41*(5), 344–358. doi:10.1016/j.wocn.2013.03.006

McCarthy, K. M., Mahon, M., Rosen, S., & Evans, B. G. (2014). Speech perception and production by sequential bilingual children: A longitudinal study of voice onset time acquisition. *Child Development, 85*(5), 1965–1980. doi:10.1111/cdev.12275

Munson, B. (2010). Levels of phonological abstraction and knowledge of socially motivated speech-sound variation: A review, a proposal, and a commentary on the Papers by Clopper, Pierrehumbert, and Tamati; Drager; Foulkes; Mack; and Smith, Hall, and Munson. *Laboratory Phonology, 1*(1), 157–177. doi:10.1515/labphon.2010.008

Nathan, L., Wells, B., & Donlan, C. (1998). Children's comprehension of unfamiliar regional accents: A preliminary investigation. *Journal of Child Language, 25*(2), 343–365. doi:10.1017/S0305000998003444

Niedzielski, N. A. (1999). The effect of social information on the perception of sociolinguistic variables. *Journal of Language and Social Psychology, 18*(1), 62–85. doi:10.1177/0261927X99018001005

Norris, D., McQueen, J. M., & Cutler, A. (2003). Perceptual learning in speech. *Cognitive Psychology, 47*(2), 204–238. doi:10.1016/S0010-0285(03)00006-9

Office for National Statistics; National Records of Scotland; Northern Ireland Statistics and Research Agency (2016). *2011 census aggregate data.* UK Data Service (Edition: June 2016).

Pierrehumbert, J. B. (2003). Phonetic diversity, statistical learning, and acquisition of phonology. *Language and Speech, 46*(Pt 2-3), 115–154. doi:10.1177/00238309030460020501

Pierrehumbert, J.B., & Gross, P. (2003). *Community phonology.* Paper presented at the Chicago Linguistic Society, Chicago, April 10-12. Retrieved from: http://babel.ling.northwestern.edu/_jbp/cls03.pdf

Pinet, M., Gan, Y., Evans, B. G., & Iverson, P. (2015). *Intelligibility of British English accents in noise for second-language learners.* Proceedings of the International Congress of Phonetic Sciences, Glasgow, UK. Retrieved from: https://www.internationalphoneticassociation.org.

Preston, D. R. (1989). *Perceptual dialectology.* Dordrecht: Foris. doi:10.1515/9783110871913

R Core Team (2018). *R: A language and environment for statistical computing.* Vienna, Austria: R Foundation for Statistical Computing. Retrieved from: http://www.R-project.org/

Scott, D. R., & Cutler, A. (1984). Segmental phonology and the perception of syntactic structure. *Journal of Verbal Learning and Verbal Behavior, 23*(4), 450–466. doi:10.1016/S0022-5371(84)90291-3

Shaw, J., Best, C. T., Docherty, G. J., Evans, B. G., Foulkes, P., Hay, J., & Mulak, K. E. (2018). Resilience of English vowel perception across regional accent variation. *Laboratory Phonology, 9*(1), 11. doi:10.5334/labphon.87

Strand, E. (1999). Uncovering the roles of gender stereotypes in speech perception. *Journal of Language and Social Psychology, 18*(1), 86–99. doi:10.1177/0261927X99018001006

Sundara, M., & Scuttellaro, A. (2011). Rhythmic difference between languages affects the development of speech perception in bilingual infants. *Journal of Phonetics, 39*(4), 505–513. doi:10.1016/j.wocn.2010.08.006

Trudgill, P., & Hannah, J. (2002). *International English: a guide to varieties of standard English* (4th ed., pp. 136–138). London, UK: Arnold.

Vertovec, S. (2007). Super-diversity and its implications. *Ethnic and Racial Studies, 30*(6), 1024–1054. doi:10.1080/01419870701599465

Wagner, L., Clopper, C. G., & Pate, J. K. (2014). Children's perception of dialect variation. *Journal of Child Language, 41*(5), 1062–1084. doi:10.1017/S0305000913000330

Weatherhead, D., White, K. S., & Friedman, O. (2016). Where are you from? Preschoolers infer background from accent. *Journal of Experimental Child Psychology, 143*, 171–178. doi:10.1016/j.jecp.2015.10.011

Wells, J. C. (1982). *Accents of English.* Cambridge: Cambridge University Press. doi:10.1017/CBO9780511611759

Werker, J. F., & Tees, R. C. (1984). Cross-language speech perception: Evidence for perceptual reorganization during the first year of life. *Infant Behavior and Development, 7*(1), 49–63. doi:10.1016/S0163-6383(84)80022-3

Xu Rattansone, N., & Demuth, K. (2013). The acquisition of coda consonants by Mandarin early child L2 learners of English. *Bilingualism: Language and Cognition, 17*(03), 646–659. doi:10.1017/S1366728913000618

Yelland, G. W., Pollard, J., & Mercuri, A. (1993). The metalinguistic benefits of limited contact with a second language. *Applied Psycholinguistics, 14*(04), 423–444. doi:10.1017/S0142716400010687

Research Article

Phonetica 2019;76:163–178
DOI: 10.1159/000493982

Received: April 3, 2017
Accepted: September 21, 2018
Published online: May 21, 2019

Subphonemic Variation and Lexical Processing: Social and Stylistic Factors

Zack Jones Cynthia G. Clopper

Ohio State University, Columbus, OH, USA

Abstract

Different pronunciation variants of the same word can facilitate lexical access, but they may be more or less effective primes depending on their phonological form, stylistic appropriateness, familiarity, and social prestige, suggesting that multiple phonological variants are encoded in the lexicon with varying strength. The current study investigated how subphonemic variation is encoded using a lexical decision task with cross-modal form priming. The results revealed that the magnitude of priming was mediated by stylistic and social properties of the auditory primes, including speaking style, talker dialect, and duration. These interactions provide evidence that phonetically reduced forms and forms that are not socially prestigious are not as robustly encoded in the lexicon as canonical forms and forms produced in prestigious varieties.

Introduction

Spoken word recognition hinges on a listener's ability to identify discrete lexical items from the continuous speech signal. Considering the high degree of acoustic variability in speech, this task of mapping variable acoustic tokens to discrete lexical representations presents a substantial processing challenge. Structural variation in the phonology of a language poses a similar challenge, in which listeners must map multiple surface realizations of a word form to its single underlying representation. For example, in American English, words that contain an intervocalic word-medial /t/ (e.g., *letter*) are often realized with a word-medial flap, rather than a stop. Despite these two phonological variants for the /t/ in *letter*, listeners identify words with these

KARGER

E-Mail karger@karger.com
www.karger.com/pho

Zack Jones
Oxley Hall 100
1712 Neil Avenue
Columbus, OH 43210 (USA)
E-Mail jones.5028@osu.edu

alternating consonants with little difficulty. The speed of lexical access, however, has been shown to vary with different phonological variants of the same word, suggesting that variants are encoded in the lexicon with differing strengths (Connine, Ranbom, & Patterson, 2008; McLennan, Luce, & Charles-Luce, 2003, 2005; Pitt, 2009; Ranbom & Connine, 2007). This evidence for the encoding of multiple phonological variants with different encoding strengths leads to the expectation that different phonetic realizations of the same word may also be encoded in the lexicon with varying strengths. The current study explored this prediction with phonetic variation arising from the combination of stylistic and social sources.

Variant Phonological Form Processing

Results from lexical processing experiments with short-term form priming have shown that different phonological variants of the same lexical item exhibit asymmetric effects, such that canonical or hyperarticulated forms (e.g., the English word *letter* produced with a word-medial [t]) facilitate lexical access relative to noncanonical or more casually produced forms (e.g., the word *letter* produced with a word-medial [ɾ]; Ernestus & Baayen, 2007; Pitt, 2009; Ranbom & Connine, 2007; Ranbom, Connine, & Yudman, 2009; Tucker, 2011; see also Gow, 2001, and McLennan et al., 2003, for similar effects in long-distance form priming and Sumner & Samuel, 2005, for similar effects in short-term semantic priming). A corpus analysis by Ranbom and Connine (2007) showed that variants with nasal flaps such as [splɪɾ̃ɚ] for the English word *splinter* are more frequent in natural speech than variants with alveolar stops (i.e., [splɪntɚ]), yet in a lexical decision task with cross-modal short-term form priming, greater facilitation was observed when visual targets were preceded by auditory primes that contained [nt] versus primes that contained [ɾ̃], demonstrating easier lexical processing for the stop variant relative to the flap variant.[1] In another study, also using a cross-modal lexical decision task with short-term form priming, Tucker and Warner (2007; see also Tucker, 2011) observed an advantage for hyperarticulated forms over reduced forms, such that tokens with fully released word-medial alveolar stops were better primes for matching visual targets than tokens with word-medial flaps (e.g., [pʌdəl] vs. [pʌɾəl] for the English target word *puddle*). These results suggest that canonical phonological forms are more robustly encoded in the lexicon than noncanonical forms.

Another source of asymmetric processing is the appropriateness of different variant forms across contexts. For example, although Ranbom and Connine (2007) observed a general cross-modal form-priming advantage for targets preceded by [nt] variants over targets preceded by [ɾ̃] variants, in an experiment in which they presented tokens in a sentential context, this advantage for the canonical form disappeared. When primes were presented within a context that strongly biased the target word semantically, the priming effect for [nt] and [ɾ̃] tokens was equivalent, suggesting that appropriate semantic context can eliminate the processing benefit for canonical forms (see also Janse, Nooteboom, & Quené, 2007, for similar effects due to appropriate phonological context). In addition, using a lexical decision task with cross-modal short-term semantic priming, Sumner (2013) manipulated the

1 An alternative interpretation of Ranbom and Connine's (2007) results is that the different numbers of segments in the stimuli containing [nt] and [ɾ̃] variants led to the observed differences in lexical processing.

stylistic context in which prime words with word-medial [nt] and [r̃] were presented to listeners. The results showed a processing benefit for trials with variant forms consistent with their stylistic context (i.e., [nt] in a carefully produced frame and [r̃] in a casually produced frame), but not for trials with a stylistic mismatch (i.e., [r̃] within a carefully produced frame). These results suggest that listeners encode not only abstract lexical items and their phonological components in the lexicon, but also contextual and stylistic information appropriate to different phonological variants.

Listener familiarity with socially meaningful phonological variants also leads to asymmetric form priming in lexical decision tasks (Sumner & Samuel, 2009; see also Rae & Warren, 2002, and Warren, Hay, & Thomas, 2007, for familiarity effects on semantic priming). Sumner and Samuel (2009) investigated the effects on cross-modal lexical form priming of listener familiarity with socially linked rhoticity in word-final syllables in English (e.g., [beɪkə] vs. [beɪkɚ] for the word *baker*). They compared three groups of listeners: rhotic speakers of General American English, rhotic speakers of New York City English, and nonrhotic speakers of New York City English. For the rhotic General American listeners who were less familiar with the word-final nonrhotic forms in New York City English, form priming was only observed with rhotic targets. For both rhotic and nonrhotic listeners from New York City, however, rhotic as well as nonrhotic targets were primed by tokens matching in form, suggesting that experience with phonological alternations related to social variation leads to stronger encoding of this variation in the lexicon, regardless of a listener's own production patterns.

In addition to listener familiarity with socially relevant phonological variation, the social prestige of particular phonological variants also impacts lexical access (Dufour, Nguyen, & Frauenfelder, 2007; Sumner & Kataoka, 2013). For example, Sumner and Kataoka (2013) presented General American listeners with three types of auditory primes in a lexical decision task with cross-modal short-term semantic priming: word-final rhotic General American English tokens, nonrhotic southern British English tokens, and nonrhotic New York City English tokens. Only the relatively prestigious General American and British English primes facilitated lexical decisions to related targets, whereas the nonstandard, nonrhotic New York City English primes did not show facilitation. This result suggests that socially prestigious forms are more robustly encoded than socially less prestigious variants (see also Sumner, Kim, King, & McGowan, 2014).

Variant Subphonemic Form Processing

While stylistic and social variation can be realized through phonological alternations, and as previous work demonstrates, these alternations are stored with listeners' representations of words, much of the stylistic and social variation that listeners encounter is realized as gradient effects on the acoustic-phonetic realization of words and their sublexical components. These gradient effects have also been shown in some previous work to affect the magnitude of priming in lexical processing tasks. For example, a short-term semantic priming study by Van de Ven, Tucker, and Ernestus (2011) looked at the effects of phonetic reduction on lexical access. Phonetically unreduced and reduced stimulus materials were elicited by asking a talker to produce speech using two different speech rates, one in which the talker was asked to produce a careful, clear style and the other in which the talker was asked to produce a faster rate. Priming was only observed when primes were phonetically unreduced, regardless of the degree of reduction in the corresponding targets, which provides evidence that phonetically unreduced forms activate semantic information related to that form, whereas phonetically reduced forms produce little to no semantic activation. This result contrasts with Sumner's (2013) results showing that both unreduced and reduced forms serve as effective semantic primes as long as the reduction is consistent with the local speaking style. However, Sumner (2013) investigated a categorical alternation, whereas Van de Ven et al. (2011) investigated effects of gradient reduction. Thus, the difference in results across the two studies may reflect the level of representation at which the reduction operates (i.e., phonemic vs. phonetic).

With respect to social variation, Clopper, Tamati, and Pierrehumbert (2016) found that listeners from both the Midland and Northern dialect regions of American English exhibited less long-distance form priming for Northern words than Midland words. Their results suggest that the encoding of Midland word tokens is stronger than the encoding of Northern word tokens, regardless of listener dialect. Given that a listener's native dialect presupposes a greater degree of familiarity with forms produced in that dialect, this result provides evidence that encoding of subphonemic dialect variation can be largely driven by factors other than familiarity. Instead, a relatively greater degree of social prestige may be tied to Midland speech than Northern speech (Clopper & Bradlow, 2008), leading to the robust encoding of Midland tokens for all listeners, regardless of dialect background (Sumner et al., 2014). Thus, similar to the encoding of phonological social variants, both listeners' familiarity with particular varieties and the social prestige they assign to particular varieties may be relevant for the encoding of subphonemic social variation.

Theoretical models based on exemplar-theoretic assumptions have been offered to explain how this fine-grained phonetic variation due to stylistic and social variation is stored in the lexicon (Johnson, 1997, 2006; Sóskuthy, 2015; Sumner et al., 2014). These models operate under a framework in which nonlinguistic variability contained in the speech signal (e.g., stylistic variation, sociophonetic variation) is processed alongside predictable linguistic variability (e.g., phonological alternations) and is furthermore encoded in a listener's lexicon as individual experienced instances that together comprise generalized abstractions that equate to lexical representations. Given these assumptions, variation arising from different sources may be encoded with different strengths, depending on such factors as exposure frequency, sociolinguistic status, etc. This differential encoding of variants, in turn, leads to the observed asymmetries in lexical processing across variants. Still, further experimental work in this domain is needed to identify how these sources of variation influence lexical encoding individually and, especially, in combination.

The Current Study

The goal of the current study was to understand how lexical encoding is influenced by stylistic and social sources of subphonemic variation. Rather than investigating one or two specific phonological variants, we considered subphonemic variation arising from multiple stylistic and social sources that differentiates word tokens produced in different contexts and by different talkers. Previous work has shown that stylistic and social sources of variation are encoded in listeners' lexical representations and lead to various patterns of form and semantic priming (Clopper et al., 2016;

Sumner, 2013; Sumner & Kataoka, 2013; Sumner & Samuel, 2009; Van de Ven et al., 2011). We therefore expected that different acoustic-phonetic realizations of words would lead to different degrees of form priming in the current study, thereby providing further evidence that gradient phonetic information related to style and social categories may be variably lexically encoded. The factors we investigated were talker dialect, talker gender, and speaking style, which have not been considered together in previous work. An orthogonal manipulation of these three factors therefore allowed us to consider the simultaneous, and potentially interacting, effects of talker dialect, talker gender, and speaking style on lexical encoding and representation. To assess lexical encoding strength, we used a cross-modal form priming paradigm with auditory primes and visual targets. A task with visual targets was selected to avoid potential misidentification of auditory target words as nonwords, which may result from the stylistic and social variation in the auditory stimulus set (Clopper & Walker, 2017).

We predicted that clear speech forms, which are less phonetically reduced than plain speech forms, would show greater priming than the more reduced plain speech forms, consistent with previous work showing advantages for canonical, unreduced, and hyperarticulated forms (Pitt, 2009; Ranbom & Connine, 2007; Tucker, 2011; Van de Ven et al., 2011). We also expected social variants that are more familiar to listeners and/or that are recognized by listeners as more socially prestigious to show more priming than less familiar and less prestigious forms (Sumner & Kataoka, 2013; Sumner et al., 2014; Sumner & Samuel, 2009). Given that formal styles tend to co-occur with fewer dialect features than less formal styles (Labov, 1972), we also expected to observe an interaction between style and talker dialect, such that the effect of talker dialect on form priming would be larger in plain speech than in clear speech. Given that female talkers tend to produce both clearer, more intelligible speech than male talkers (Bradlow, Torretta, & Pisoni, 1996) and more prestigious variants than male talkers (Labov, 1990), we expected tokens produced by female talkers to show greater priming than tokens produced by male talkers, especially for the less familiar dialect and the plain speaking style. Thus, we also expected to observe interactions between talker gender, talker dialect, and style.

In addition to manipulating stylistic and social properties of the auditory primes, we also controlled for two properties of individual lexical items that affect listeners' ability to access appropriate lexical representations: lexical frequency and phonological neighborhood density (e.g., Balota & Chumbley, 1984; Clopper & Walker, 2017; Goldinger, Luce, & Pisoni, 1989; Luce & Pisoni, 1998). Across priming paradigms, low-frequency words show more priming than high-frequency words (Becker, 1979; Colombo, 1986; Van de Ven et al., 2011; Van Petten & Kutas, 1990). In form-priming tasks, both facilitation for phonologically similar primes and targets, as well as inhibition for similar primes and targets have been observed, consistent with lexical activation and competition of similar words (Clopper & Walker, 2017; Goldinger, Luce, Pisoni, & Marcario, 1992; Hillinger, 1980; Radeau, Morais, & Dewier, 1989; Slowiaczek & Pisoni, 1986). We expected to observe greater priming for trials that featured low-frequency primes compared to trials that featured high-frequency primes and more priming for trials with low-density primes compared to trials with high-density primes, given the greater degree of lexical competition for high-density words than low-density words.

Phonetica 2019;76:163–178
DOI: 10.1159/000493982

Table 1. Examples of the three different trial types presented in the lexical decision task

Trial type	Auditory prime token	Visual target word	Trials, *n*
Matching	*dress*	DRESS	78 (25%)
Unrelated	*thief*	OOZE	78 (25%)
Nonword	*bus*	NODGE	156 (50%)

Methods

Participants

Ninety monolingual lifetime residents (29 male, 61 female) of the Midland dialect region of American English participated in the current study. Regional dialect background was determined using the geographic descriptions given by Labov, Ash, and Boberg (2005). The average age of participants was 20.9 years (range: 17–49 years). An additional 145 participants were tested, but their data were not included in the analysis because they were multilingual ($n = 69$), from a dialect region other than the Midland region ($n = 72$), reported a history of hearing and/or language impairment ($n = 3$), or failed to complete the task as instructed ($n = 1$). All participants received partial course credit in an undergraduate linguistics course for their participation.

Stimulus Materials

The cross-modal lexical decision task in the current study included two sets of stimulus materials: a set of auditory word tokens, presented as primes, and a set of visual word and nonword letter strings, presented as targets. The stimulus materials were presented in three types of trials: matching trials in which the auditory prime and the visual target were the same word type, unrelated trials in which the auditory prime and the visual target were different word types, and nonword trials in which the auditory prime was a real word of English and the visual target was a nonword. Examples of matching, unrelated, and nonword trials are shown in Table 1.

The auditory primes were extracted from recorded read passages in the Ohio State Stories Corpus (Burdin, Turnbull, & Clopper, 2015). These passages were elicited twice from each talker in the corpus: first in a plain lab speaking style and then in a clear lab speaking style. Plain lab speech was elicited by instructing talkers to speak as if they were addressing a close friend or family member. Clear lab speech was elicited by instructing talkers to speak as if they were addressing a nonnative or hearing-impaired listener. Acoustic analysis of the corpus materials confirmed that plain speech was produced with significantly shorter and less dispersed vowels than clear speech, as expected (Burdin et al., 2015). The speech of 4 male and 4 female adult speakers each from the Midland and Northern dialects of American English were used for the current study.

The passages in the corpus were designed to include a total of 234 target English word types (232 monosyllabic, 2 disyllabic).[2] These target words varied orthogonally in lexical frequency and phonological neighborhood density, as defined by Nusbaum, Pisoni, and Davis (1984). Of the 234 target lexical items, one third ($n = 78$) served as the auditory primes and visual targets for the matching trials, in which the auditory prime and the visual target matched for lexical identity. Of the remaining target lexical items, half ($n = 78$) served as auditory primes and half ($n = 78$) as visual targets for the unrelated trials, in which the prime and the target were unrelated both semantically and phonologically. For the unrelated trials, target words were deemed phonologically unrelated if they shared no common segments in their onsets, nuclei, or codas. Semantic unrelatedness of target words was determined impressionistically by the first author. Three sepa-

[2] In the original corpus, two disyllabic word targets were used to balance the set of words for lexical frequency and neighborhood density. To maintain this balance, all 234 target words were used in the current study.

rate lists of unique auditory tokens of the 234 target items were created to reduce the impact of specific tokens on performance. The auditory tokens for the matching and unrelated trials in each of the lists were balanced for seven factors: speaking style (clear vs. plain), talker dialect (Midland vs. North), talker gender (male vs. female), lexical frequency (high vs. low), phonological neighborhood density (high vs. low), semantic predictability (high vs. low), and discourse mention (first vs. second). They were also balanced for individual talker and target word stressed vowel, which consisted of one of six vowels: /u, i, ɔ, ɛ, ɑ, æ/. Target word types were counterbalanced across matching and unrelated trials within each of the three lists.

In addition to the 78 matching trials and 78 unrelated trials, 156 trials consisted of auditory filler primes and nonword targets, for a total of 312 trials. The auditory primes for these nonword trials were 156 different words extracted from the same read passages as the target primes. The 156 filler items were also balanced for all of the factors described above. Nonword targets were created to resemble English words in their syllable structure and phonotactic patterns. Of the 156 nonwords, 153 were monosyllabic and 3 were disyllabic to roughly mirror the syllabic makeup of the 234 word types presented in the matching and unrelated trials. The same set of 156 filler word primes and nonword targets was used in all three lists. The complete list of word primes and word and nonword targets are shown in Table A1 of the Appendix.

Procedure

Participants were seated at a personal computer equipped with circumaural headphones and a five-alternative button box. The experiment was conducted using SuperLab 4.5 (Cedrus Corporation, San Pedro, CA, USA). On each trial of the experiment, participants heard an auditory prime over the headphones, which was immediately followed by the appearance of a visual letter string target on the computer screen (i.e., with a 0-ms interstimulus interval between the offset of the auditory prime and the onset of the visual target). Visual stimuli were presented in the center of the computer screen in black 72-point Arial font against a white background. Participants were instructed to use one of two labeled buttons on the button box to indicate whether the visual letter string was or was not a real word of English. "Word" responses were made using the rightmost button on the button box, and "nonword" responses were made using the leftmost button on the button box.[3] Participants were asked to respond as quickly as possible while maintaining a high level of accuracy. All trials were randomized separately for each participant and presented in three blocks of 104 trials each with self-paced breaks between blocks.

The equal distribution of word and nonword trials, the low proportion of matching trials (25%), and the 0-ms interstimulus interval were expected to limit strategic response biases so that the effects of style, talker gender, and talker dialect on lexical access could be assessed (Hamburger & Slowiaczek, 1996). Additionally, the careful balancing of style, talker gender, and talker dialect across primes in the matching and unrelated trials should ensure that any remaining effects of response strategy on the magnitude of priming were distributed equally across the critical factors in the design.

Statistical Analysis

Only response times to correct matching and unrelated real word trials were analyzed statistically. Data from 811 incorrect trials (5.8% of all trials; 3.4% of matching trials and 8.1% of unrelated trials) were therefore excluded. A series of data-pruning steps was then taken to minimize effects of outliers and to better fulfill analysis assumptions of a normally distributed data set. First, data from 29 correct trials (<1% of all correct trials) with raw response times shorter than 200 ms or greater than 2,500 ms were excluded as trials during which listeners may have lost attention to the task. Second, response times were log-transformed to achieve a data distribution that more closely approached normality. Third, 629 trials (4.8% of all correct trials; 2.3% of matching trials and 2.5% of unrelated trials) were excluded by removing trials with log-trans-

[3] Hand dominance was not assessed because only correct word responses were analyzed. Any differences in response times to word responses as a function of hand dominance were captured by the random intercepts for participants in the statistical analysis (see the section "Statistical Analysis").

DOI: 10.1159/000493982

formed response times that were greater than or less than 2 standard deviations from each individual listener's mean log-transformed response time to reduce the effects of outliers. Although the statistical analysis involved log-transformed response times as the dependent variable, raw response times, rather than log-transformed times or model estimates, are shown in the figures and tables in the Results section to facilitate interpretation.

A linear mixed-effects regression model was built to assess how factors associated with the prime words influenced lexical priming on matching trials. A priming effect is observed in this experimental paradigm when response times to matching trials are significantly faster than response times to unrelated trials. The magnitude of this effect may be enhanced or attenuated depending on attributes of the auditory primes (Radeau, Morais, & Segui, 1995). As an overall priming effect was expected for response times to matching trials, of critical interest in the statistical analysis was the presence or absence of interactions between trial type (matching vs. unrelated) and the other experimental factors related to the prime words.

To investigate how social and stylistic factors influence the magnitude of cross-modal priming, interactions with three factors related to the auditory primes were analyzed: prime talker gender (female or male), talker dialect (Midland or North), and speaking style (clear or plain). Thus, the fixed effects in the model were talker gender, talker dialect, speaking style, and their interactions with each other and trial type (i.e., trial type × talker gender × talker dialect × speaking style). Log-transformed lexical frequency and phonological neighborhood density of the auditory primes were expected to mediate the priming effect and thus interact with trial type. The two lexical factors and their separate interactions with trial type (i.e., trial type × lexical frequency + trial type × phonological neighborhood density) were therefore included as covariates. Finally, although vowel duration was not directly manipulated in the current study, the duration of the auditory primes' stressed vowel was also included as a covariate in the model, along with its interaction with trial type (i.e., trial type × duration) to account for natural vowel duration variation in the corpus materials. Although semantic predictability and discourse mention were manipulated in the corpus and balanced in the stimulus materials, these factors were not included in our analysis because the stimuli consisted of single-word tokens and thus were presented without their original discourse context. Given the effects of these two factors on vowel duration (Bard, Anderson, Sotillo, Aylett, Doherty-Sneddon, & Newlands, 2000; Clopper & Pierrehumbert, 2008; Fowler & Housum, 1987; Lieberman, 1963), however, the effects of these factors on lexical priming were captured by the vowel duration covariate and its interaction with trial type.

All binary variables (i.e., trial type, talker gender, talker dialect, and speaking style) were sum-contrast coded, and all continuous variables (i.e., lexical frequency, neighborhood density, and vowel duration) were centered for analysis. A maximal random effects structure was initially specified, and random slopes were removed until model convergence was achieved. In the final model, random intercepts for prime word type, target word type, and participant were included, as well as a random slope for trial type by prime word and random slopes for trial type and talker dialect by participant. No random slopes by target word type were included. To determine the statistical significance of the main effects and interactions in the model, a series of log-likelihood comparisons was conducted between models lacking a factor or interaction of interest and a more fully specified model that included the factor or interaction, resulting in model comparisons that always differed by one parameter. The full model output is included in Table A2 of the Appendix.

Results

The log-likelihood comparisons revealed three significant interactions, all of which involved trial type, demonstrating that a number of factors significantly mediated the magnitude of priming. The main effects and interactions that were significant in the log-likelihood comparisons are summarized in Table 2. First, the three-way interaction among trial type, speaking style, and talker dialect was significant, with $\chi^2 = 3.95$, df = 1, and $p < 0.05$. While a priming effect was observed for all trials,

Table 2. Significant main effects and interactions predicting log-transformed response times

Fixed effects	Estimate	*t* value
Trial type	−0.053	−20.5
Word frequency	0.006	2.3
Trial type × word phonological density	<0.001	2.6
Trial type × vowel duration	−0.136	−3.5
Trial type × speaking style	−0.003	−1.5
Trial type × speaking style × talker dialect	−0.004	−2.0

Table 3. By-target mean response times for each trial type, speaking style, and talker dialect

	Clear style		Plain style	
	Midland	Northern	Midland	Northern
Unrelated	588 (72)	581 (79)	574 (66)	585 (70)
Matching	515 (52)	520 (52)	523 (52)	524 (49)
Priming	73	61	51	61

Means are given in milliseconds, and standard deviations are shown in parentheses. The magnitude of priming is the difference between matching and unrelated trials.

such that matching trials were responded to more quickly than unrelated trials overall, this three-way interaction shows that the magnitude of this effect was mediated by dialect and style. In particular, as shown in Table 3, priming was strongest for the clear Midland primes and weakest for the plain Midland primes. Priming for the clear and plain Northern primes was intermediate.

To confirm this interpretation of the three-way interaction, four linear mixed-effects models were built using treatment contrast coding for trial type, style, and talker dialect with reference levels for each style by talker dialect combination. The reference level for trial type was unrelated trials for all four post hoc models. The simple effect of trial type was significant, $|\beta| > 0.09$, $|t| > 2$ in all four models, confirming significant priming across styles and talker dialects. In addition, the main effects of style and talker dialect were not significant in any of the four models, confirming that response times to unrelated trials were similar across style and talker dialect. The two models featuring Midland primes as the reference level showed a significant two-way interaction of trial type and style, $|\beta| = 0.025$, $|t| = 2.5$, confirming a significant difference in priming between clear and plain Midland primes. The two models featuring Northern primes as the reference level showed a nonsignificant two-way interaction between trial type and talker dialect, confirming the lack of difference in priming between clear and plain Northern primes. None of the models revealed a significant two-way interaction between trial type and talker dialect, confirming the intermediate magnitude of priming for Northern primes overall. The greater magnitude of priming when Midland primes were produced in clear speech than plain

Phonetica 2019;76:163–178
DOI: 10.1159/000493982

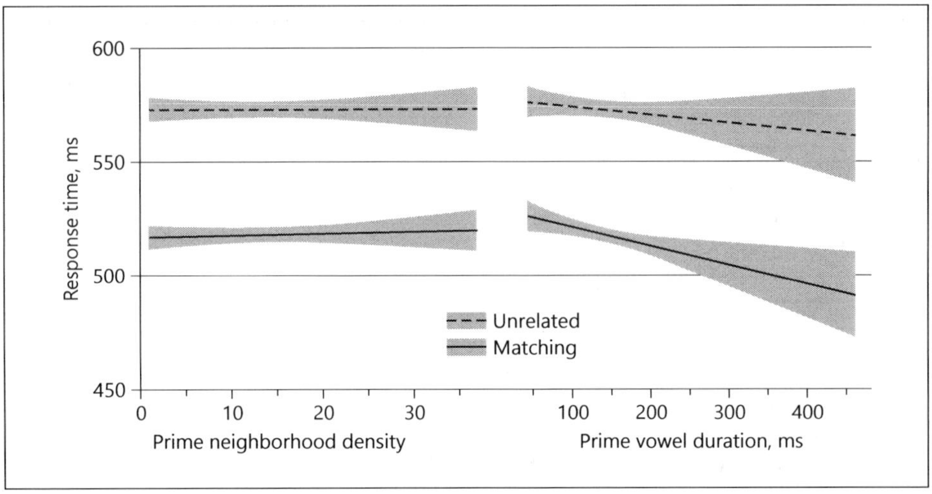

Fig. 1. Linear model predictions showing two significant two-way interactions. The trial type × prime word phonological neighborhood density interaction is illustrated on the left and the trial type × prime token vowel duration interaction is illustrated on the right.

speech suggests that hyperarticulated forms lead to faster lexical processing compared to reduced forms, but this advantage for hyperarticulated forms is limited to highly familiar or more standard dialects.

Second, the two-way interaction between trial type and prime phonological neighborhood density was also significant, $\chi^2 = 6.90$, df = 1, $p < 0.01$. This interaction is illustrated in the left panel of Figure 1. A greater priming effect was observed for primes with few phonological neighbors compared to primes with many neighbors. This result suggests that the lexical activation of the few phonological neighbors of a low-density prime word allows the matching target to be processed quickly. In contrast, the greater overall lexical activation of the many neighbors of a high-density prime word slows recognition of the matching target. That is, words with few neighbors benefit more from the corroborating evidence between a matching prime and target than high-density words.

Third, a significant trial type x prime vowel duration interaction was observed, with $\chi^2 = 12.10$, df = 1, and $p < 0.001$. The right panel of Figure 1 shows this two-way interaction, which reveals a greater priming effect for trials that featured auditory primes with relatively long vowels compared to trials that featured primes with relatively short vowels. This interaction suggests that when primes and targets match, phonetically unreduced (i.e., longer) forms show a processing advantage over reduced (i.e., shorter) forms, but this advantage for unreduced forms relative to reduced forms is not observed when primes and targets are unrelated.[4] The increased priming effect for unreduced tokens provides further evidence that listeners' lexical represen-

[4] Although other factors, including intrinsic vowel duration differences and phonetic context, could also play a role in vowel duration variation, these word-specific effects are captured by the random intercepts and trial type slopes for prime word.

tations of isolated words more closely resemble canonical forms of lexical items, as has been suggested in previous work (Sumner, 2013).

Lastly, a significant main effect of prime lexical frequency was observed, with $\chi^2 = 4.52$, df = 1, and $p < 0.05$, such that trials that included highly frequent prime words were responded to more slowly overall than trials that included less frequent prime words. This main effect, which operated independently of trial type, suggests that when presented with a frequent prime, strong lexical activation of the prime word may interfere with access to the visual target, leading to slower response times overall.

Discussion and Conclusion

The significant trial type × speaking style × talker dialect interaction revealed more priming on matching trials with clear Midland primes relative to trials with plain Midland primes. Matching trials with clear and plain Northern primes exhibited an intermediate degree of priming. These results provide evidence that clear Midland word tokens are more robustly encoded in the lexicon than plain Midland tokens, whereas clear and plain Northern tokens are not differentially encoded. The result for the Midland tokens is consistent with previous work showing that word tokens produced in a clear style and exhibiting little phonetic reduction are encoded in listeners' representations of isolated words (Pitt, 2009; Ranbom & Connine, 2007; Tucker, 2011; Van de Ven et al., 2011). However, the interaction between trial type, speaking style, and talker dialect also suggests that the encoding of stylistic variation and social variation is not independent. Given that clear speech is expected to exhibit fewer marked dialect features than plain speech (Labov, 1972), a larger dialect difference in priming was expected for plain speech than clear speech. However, a different pattern was observed in the current study, such that a speaking style effect was observed only for the native Midland dialect and no robust effects of talker dialect were observed for either speaking style. Thus, the results suggest that representations of clear word tokens are more robustly encoded than representations of plain or casual word tokens, but only when the variants are reasonably familiar to listeners and/or are produced in a variety that is perceived as more prestigious or more similar to a standard variety (cf. Sumner et al., 2014). Unlike talker regional dialect and speaking style, talker gender did not significantly affect lexical encoding in this task. However, vowel realization is not predicted to differ substantially in systematic ways across talker genders, which may explain this lack of evidence for talker gender effects.

The intermediate results for the Northern tokens provide a contrast to previous work showing that dialect variants that are more familiar and/or recognized as more socially prestigious are more robustly encoded than variants produced in other varieties (Dufour et al., 2007; Sumner & Kataoka, 2013; Sumner & Samuel, 2009). All of the listeners in the current study came from the Midland dialect region and were thus presumably more familiar with Midland speech than Northern speech. Midland speech has also been suggested to be closer to a conceptual standard American English variety than Northern speech (Clopper & Bradlow, 2008; Clopper, Pierrehumbert, & Tamati, 2010). Given that Midland speech is both more familiar and likely viewed as more prestigious or standard by our listeners, Northern tokens were expected to exhibit less priming than Midland tokens, especially in the plain style. Although numerically less priming was observed for clear Northern primes than clear Midland primes, this difference was not significant, and in plain speech, numerically more priming was observed for Northern primes than Midland primes (Table 2). Further work is therefore needed to assess how familiarity and perceived prestige mediate the encoding of stylistic variation in listeners' representations of isolated words in a nonlocal dialect.

The observed interplay between speaking style and regional dialect information in lexical encoding lends further support to models of word representation that are based on experienced exemplars (Goldinger, 1998; Johnson, 1997, 2006; Pierrehumbert, 2001; Sóskuthy, 2015; Sumner et al., 2014). In addition, these results suggest that listeners do not only store disparate talker-related and stylistically relevant information with the linguistic content of words they encounter. Rather, the stylistic and social information linked to listeners' stored representations of speech tokens are in turn linked to one another. Importantly, these different types of information in the speech signal are encoded collectively, creating relationships in the lexicon across styles, social categories, and phonetic variation that together mediate lexical access, as observed in the interactions in the current study. Thus, from an exemplar-theoretic perspective, individual word tokens are linked to higher-level abstract linguistic, stylistic, and social categories, which are themselves interconnected.

These proposed relationships among linguistic, stylistic, and social categories are consistent with Sumner et al.'s (2014) model of socially weighted lexical encoding, in which spoken word tokens are stored in the lexicon according to their socially relevant phonetic characteristics. Given previous results such as those discussed in the section "Variant Phonological Form Processing" that show asymmetrical processing of different pronunciation variants, Sumner et al.'s (2014) model predicts differential encoding of word tokens depending on their sociophonetic properties. It also predicts the interactive encoding and long-term storage of word tokens and their sociophonetic characteristics in the lexicon. Our finding that more priming is observed for clear Midland tokens than for plain Midland tokens provides strong support for Sumner et al.'s (2014) claim that socially idealized forms are more robustly encoded than other forms. Our results furthermore demonstrate the interactivity that exists among the different sources of social and stylistic information that are tied to lexical representations.

Acknowledgments

We would like to thank Megan Dailey and Jessica Hanson for their contributions to stimulus preparation and Nathanael Fath for assisting with data collection. This work was partially supported by the National Science Foundation (BCS-1056409).

Disclosure Statement

The authors declare that there is no conflict of interest involved with the current study.

Appendix

Table A1

List of the 234 word types presented as auditory primes and visual targets in matching and unrelated trials, the 156 word types presented as auditory fillers in nonword trials, and the 156 nonwords presented as visual targets in nonword trials.

Target words	ask, axe, back, bad, bag, bath, beach, beak, beat, bed, beef, best, bet, black, bleed, booth, boots, boss, bought, box, brag, bread, breast, breath, brief, broth, bruise, cat, catch, caught, cause, chalk, cheap, cheese, chef, chess, chest, chief, chop, class, clause, claws, clogs, cloth, cops, cot, cough, crab, crash, crew, crop, cute, deaf, death, deck, depth, desk, dodge, dog, draft, dream, dress, drop, fact, faun, feast, feud, flaw, flax, fleas, fleece, flesh, floss, flute, food, fought, fox, fresh, frock, frog, fruit, gauze, glass, gnats, god, gone, goose, greed, grief, group, hat, haunt, head, heat, hot, huge, jaw, jazz, jeans, job, juice, June, knee, knots, law, lawn, leg, lodge, long, loose, lost, mask, match, math, mead, meat, mesh, mess, mob, mood, moon, moose, moss, moth, nap, neck, nest, net, news, niece, nude, off, ooze, ox, pact, pass, path, pet, please, pot, priest, proof, quest, raft, rat, rest, rock, root, rot, sad, sauce, scab, scheme, scrap, scratched, sect, seed, shed, sheeps, sheets, shock, shoe, shop, shot, shrew, sketch, ski, slapped, sled, sleeve, sleuth, sloth, smog, smooth, snag, snooze, soft, soothe, speech, speed, splash, squad, squeak, staff, step, stew, stock, stop, straw, stream, street, stress, strong, suit, suite, swab, swan, swat, sweat, tag, talk, task, tax, teeth, test, theft, theme, thief, thought, thread, threat, top, toss, track, treat, tree, true, truth, tube, tunic, tweed, tweet, view, walk, wash, watch, water, weed, wheat, yacht, yeast, youth, zest
Filler words	add, aid, ate, bars, bathe, bill, bird, bit, blood, bugs, build, bus, came, care, cars, charge, check, claim, clean, clothes, club, corn, costs, course, crowd, cup, curry, cuts, dance, day, deer, doors, dorm, drink, drugs, eat, end, face, false , farms, fashion, feel, feet, fell, felt, find, flies, flight, floor, friend, front, gave, ghost, give, growth, guilt, gym, hand, health, hemp, hill, home, hosts, key, lack, laid, laps, leaves, lift, mail, maize, miss, mix, mom, mountains, mouth, myths, name, need, night, noise, note, nurse, orange, pack, page, parts, peck, phone, place, plane, rage, rain, ran, reach, read, ride, right, road, room, rope, run, school, scorn, scourge, seas, secret, senate, send, serve, sew, ship, sight, silk, sink, size, skill, skin, skirt, sleep, slept, smell, smoke, snow, sorts, spare, split, spread, start, stray, sun, sway, taste, teams, think, threw, time, tips, tone, took, towns, trust, turn, twigs, use, vowed, wave, way, week, win, winter, wore, work, world, yards, year
Nonwords	baff, bave, bice, blant, blint, bresh, brints, bulp, chilm, chim, clarry, cliss, crabe, crench, dalt, dape, darf, derd, derse, dest, dight, doan, dobe, dorps, drail, dreff, drep, drig, driss, droups, dunch, famps, fance, feg, fice, firch, fisp, florn, flurot, forch, foss, frades, framp, fross, furnt, gair, garve, geat, glead, goan, goub, grat, gree, hadge, haint, hame, hammle, hudge, jate, jid, joat, jub, juff, keck, keetch, kiff, kipe, klut, kulk, labe, ladge, launt, leabs, leet, liff, loffs, loids, loise, loke, lorm, louch, lounds, lumb, marp, maved, meach, meef, mish, moton, mun, nard, nart, neak, nire, nodge, nuck, nunk, nunt, pake, pib, plam, plast, plick, plin, plut, polt, poyed, pretta, prib, prook, prug, quep, rall, ribe, rizz, sade, shope, skaze, slabe, slince, sluck, smad, snaff, snarts, sool, sout, speeze, spife, spish, spreeve, spuck, stod, stoofed, surce, tammed, tane, taw, thork, tifth, trave, trid, trine, trollet, troom, troth, trum, twack, verp, vitch, vore, vork, wace, wibb, woose, wope, yole

Table A2
Model output with log-transformed response times as dependent variable.

Fixed effects	Estimate	*t* value
Intercept	6.279	553.8
Trial type	–0.053	–20.5
Talker gender	<–0.001	–0.1
Speaking style	–0.001	–0.6
Talker dialect	<–0.001	–0.3
Word frequency	0.006	2.3
Word phonological density	<0.001	3.2
Vowel duration	–0.156	–4.1
Trial type × talker gender	<–0.001	–0.2
Trial type × speaking style	–0.003	–1.5
Trial type × talker dialect	<0.001	0.2
Talker gender × speaking style	<0.001	0.4
Talker gender × talker dialect	–0.004	–1.9
Speaking style × talker dialect	<–0.001	–0.3
Trial type × word frequency	0.002	0.7
Trial type × word phonological density	<0.001	2.6
Trial type × vowel duration	–0.136	–3.5
Trial type × talker gender × speaking style	<0.001	0.0
Trial type × talker gender × talker dialect	<0.001	0.3
Trial type × speaking style × talker dialect	–0.004	–2.0
Talker gender × speaking style × talker dialect	<0.001	0.1
Trial type × talker gender × speaking style × talker dialect	<–0.001	–0.1

References

Balota, D. A., & Chumbley, J. I. (1984). Are lexical decisions a good measure of lexical access? The role of word frequency in the neglected decision stage. *J Exp Psychol: Human, 10*(3), 340–357. doi: 10.1037/0096-1523.10.3.340

Bard, E. G., Anderson, A. H., Sotillo, C., Aylett, M., Doherty-Sneddon, G., & Newlands, A. (2000). Controlling the intelligibility of referring expressions in dialogue. *Journal of Memory and Language, 42*(1), 1–22. doi:10.1006/jmla.1999.2667

Becker, C. A. (1979). Semantic context and word frequency effects in visual word recognition. *J Exp Psychol: Human, 5*(2), 252–259. doi:10.1037/0096-1523.5.2.252

Bradlow, A. R., Torretta, G. M., & Pisoni, D. B. (1996). Intelligibility of normal speech I: Global and fine-grained acoustic-phonetic talker characteristics. *Speech Communication, 20*(3), 255–272. doi:10.1016/S0167-6393(96)00063-5

Burdin, R. S., Turnbull, R., & Clopper, C. G. (2015). Interactions among lexical and discourse characteristics in vowel production. *Proceedings of Meetings on Acoustics, 22*(060005), 1–18. doi:10.1121/2.0000084

Clopper, C. G., & Bradlow, A. R. (2008). Perception of dialect variation in noise: Intelligibility and classification. *Language and Speech, 51*(Pt 3), 175–198. doi:10.1177/0023830908098539

Clopper, C. G., & Pierrehumbert, J. B. (2008). Effects of semantic predictability and regional dialect on vowel space reduction. *The Journal of the Acoustical Society of America, 124*(3), 1682–1688. doi: 10.1121/1.2953322

Clopper, C. G., Pierrehumbert, J. B., & Tamati, T. N. (2010). Lexical neighborhoods and phonological confusability in cross-dialect word recognition in noise. *Laboratory Phonology, 1*(1), 65–92. doi: 10.1515/labphon.2010.005

Clopper, C. G., Tamati, T. N., & Pierrehumbert, J. B. (2016). Variation in the strength of lexical encoding across dialects. *Journal of Phonetics, 58*, 87–103. doi:10.1016/j.wocn.2016.06.002

Clopper, C. G., & Walker, A. (2017). Effects of lexical competition and dialect exposure on phonological priming. *Language and Speech, 60*(1), 85–109. doi:10.1177/0023830916643737

Colombo, L. (1986). Activation and inhibition with orthographically similar words. *J Exp Psychol: Human, 12*(2), 226–234. doi:10.1037/0096-1523.12.2.226

Connine, C. M., Ranbom, L. J., & Patterson, D. J. (2008). Processing variant forms in spoken word recognition: The role of variant frequency. *Perception & Psychophysics, 70*(3), 403–411. doi:10.3758/PP.70.3.403

Dufour, S., Nguyen, N., & Frauenfelder, U. H. (2007). The perception of phonemic contrasts in a non-native dialect. *The Journal of the Acoustical Society of America, 121*(4), EL131–EL136. doi.org/10.1121/1.2710742

Ernestus, M., & Baayen, R. H. (2007). The comprehension of acoustically reduced morphologically complex words: The roles of deletion, duration, and frequency of occurrence. In J. Trouvain & W. J. Barry (Eds.), *Proceedings of the 16th International Congress of Phonetic Sciences* (pp. 773–776). Dudweiler, Germany: Pirrot.

Fowler, C. A., & Housum, J. (1987). Talkers' signaling of "new" and "old" words in speech and listeners' perception and use of the distinction. *Journal of Memory and Language, 26*(5), 489–504. doi: 10.1016/0749-596X(87)90136-7

Goldinger, S. D. (1998). Echoes of echoes? An episodic theory of lexical access. *Psychological Review, 105*(2), 251–279. doi:10.1037/0033-295X.105.2.251

Goldinger, S. D., Luce, P. A., & Pisoni, D. B. (1989). Priming lexical neighbors of spoken words: Effects of competition and inhibition. *Journal of Memory and Language, 28*(5), 501–518. doi:10.1016/0749-596X(89)90009-0

Goldinger, S. D., Luce, P. A., Pisoni, D. B., & Marcario, J. K. (1992). Form-based priming in spoken word recognition: The roles of competition and bias. *J Exp Psychol: Learn, 18*(6), 1211–1238. doi: 10.1037/0278-7393.18.6.1211

Gow, D. W., Jr. (2001). Assimilation and anticipation in continuous spoken word recognition. *Journal of Memory and Language, 45*(1), 133–159. doi:10.1006/jmla.2000.2764

Hamburger, M., & Slowiaczek, L. M. (1996). Phonological priming reflects lexical competition. *Psychonomic Bulletin & Review, 3*(4), 520–525. doi:10.3758/BF03214558

Hillinger, M. L. (1980). Priming effects with phonemically similar words: The encoding-bias hypothesis reconsidered. *Memory & Cognition, 8*(2), 115–123. doi:10.3758/BF03213414

Janse, E., Nooteboom, S. G., & Quene, H. (2007). Coping with gradient forms of /t/-deletion and lexical ambiguity in spoken word recognition. *Language and Cognitive Processes, 22*(2), 161–200. doi:.1080/01690960500371024

Johnson, K. (1997). Speech perception without speaker normalization: An exemplar model. In K. Johnson & J. W. Mullennix (Eds.), *Talker Variability in Speech Processing* (pp. 145–165). San Diego: Academic Press.

Johnson, K. (2006). Resonance in an exemplar-based lexicon: The emergence of social identity and phonology. *Journal of Phonetics, 34*(4), 485–499. doi:10.1016/j.wocn.2005.08.004

Labov, W. (1972). *Sociolinguistic Patterns*. Philadelphia: University of Pennsylvania Press.

Labov, W. (1990). The intersection of sex and social class in the course of linguistic change. *Language Variation and Change, 2*(02), 205–254. doi:10.1017/S0954394500000338

Labov, W., Ash, S., & Boberg, C. (2005). *The Atlas of North American English: Phonetics, Phonology and Sound Change*. Berlin: Mouton de Gruyter. doi:10.1515/9783110167467

Lieberman, P. (1963). Some effects of semantic and grammatical context on the production and perception of speech. *Language and Speech, 6*(3), 172–187. doi:10.1177/002383096300600306

Luce, P. A., & Pisoni, D. B. (1998). Recognizing spoken words: The neighborhood activation model. *Ear and Hearing, 19*(1), 1–36. doi:10.1097/00003446-199802000-00001

McLennan, C. T., Luce, P. A., & Charles-Luce, J. (2003). Representation of lexical form. *J Exp Psychol: Learn, 29*(4), 539–553. doi:10.1037/0278-7393.29.4.539

McLennan, C. T., Luce, P. A., & Charles-Luce, J. (2005). Representation of lexical form: Evidence from studies of sublexical ambiguity. *J Exp Psychol: Human, 31*(6), 1308–1314. doi:10.1037/0096-1523.31.6.1308

Nusbaum, H. C., Pisoni, D. B., & Davis, C. K. (1984). *Sizing up the Hoosier Mental Lexicon: Measuring the familiarity of 20,000 words*. Research on Speech Perception Progress Report No. 10 (pp. 357–376). Bloomington, IN: Speech Research Laboratory, Indiana University.

Pierrehumbert, J. B. (2001). Exemplar dynamics: Word frequency, lenition and contrast. In J. Bybee & P. Hopper (Eds.), *Frequency and the Emergence of Linguistic Structure* (pp. 137–157). Amsterdam: John Benjamins. doi:10.1075/tsl.45.08pie

Pitt, M. A. (2009). How are pronunciation variants of spoken words recognized? A test of generalization to newly learned words. *Journal of Memory and Language, 61*(1), 19–36. doi:10.1016/j.jml.2009.02.005

Radeau, M., Morais, J., & Dewier, A. (1989). Phonological priming in spoken word recognition: Task effects. *Memory & Cognition, 17*(5), 525–535. doi:10.3758/BF03197074

Radeau, M., Morais, J., & Segui, J. (1995). Phonological priming between monosyllabic spoken words. *Journal of Experimental Psychology: Human, 21*, 1297–1310.

Rae, M., & Warren, P. (2002). Goldilocks and the three beers: Sound merger and recognition in NZE. *New Zealand English Journal, 16*, 33–41.

Ranbom, L. J., & Connine, C. M. (2007). Lexical representation of phonological variation in spoken word recognition. *Journal of Memory and Language, 57*(2), 273–298. doi:10.1016/j.jml.2007.04.001

Ranbom, L. J., Connine, C. M., & Yudman, E. M. (2009). Is phonological context always used to recognize variant forms in spoken word recognition? The role of variant frequency and context distribution. *Journal of Experimental Psychology: Human, 35*(4), 1205–1220. doi:10.1037/a0015022

Slowiaczek, L. M., & Pisoni, D. B. (1986). Effects of phonological similarity on priming in auditory lexical decision. *Memory & Cognition, 14*(3), 230–237. doi:10.3758/BF03197698

Sóskuthy, M. (2015). Understanding change through stability: A computational study of sound change actuation. *Lingua, 163*, 40–60. doi:10.1016/j.lingua.2015.05.010

Sumner, M. (2013). A phonetic explanation of pronunciation variant effects. *The Journal of the Acoustical Society of America, 134*(1), EL26–EL32. doi:10.1121/1.4807432

Sumner, M., & Kataoka, R. (2013). Effects of phonetically-cued talker variation on semantic encoding. *The Journal of the Acoustical Society of America*, 134(6), EL485–EL491. https://doi.org/10.1121/1.4826151

Sumner, M., Kim, S. K., King, E., & McGowan, K. B. (2014). The socially weighted encoding of spoken words: A dual-route approach to speech perception. *Frontiers in Psychology, 4*, 1015. doi:10.3389/fpsyg.2013.01015

Sumner, M., & Samuel, A. G. (2005). Perception and representation of regular variation: The case of word-final /t/. *Journal of Memory and Language, 52*(3), 330–346. doi:10.1016/j.jml.2004.11.004

Sumner, M., & Samuel, A. G. (2009). The effect of experience on the perception and representation of dialect variants. *Journal of Memory and Language, 60*(4), 487–501. doi:10.1016/j.jml.2009.01.001

Tucker, B. V. (2011). The effect of reduction on the processing of flaps and /g/ in isolated words. *Journal of Phonetics, 39*(3), 312–318. doi:10.1016/j.wocn.2010.12.001

Tucker, B. V., & Warner, N. (2007). Inhibition of processing due to the reduction of the American English flap. *Proceedings of the 16th International Congress of Phonetic Sciences* (pp. 773–776). Saarbrücken, Germany.

van de Ven, M., Tucker, B. V., & Ernestus, M. (2011). Semantic context effects in the comprehension of reduced pronunciation variants. *Memory & Cognition, 39*(7), 1301–1316. doi:10.3758/s13421-011-0103-2

Van Petten, C., & Kutas, M. (1990). Interactions between sentence context and word frequency in event-related brain potentials. *Memory & Cognition, 18*(4), 380–393. doi:10.3758/BF03197127

Warren, P., Hay, J., & Thomas, B. (2007). The loci of sound change effects in recognition and perception. In J. Cole & J. I. Hualde (Eds.), *Laboratory Phonology 9* (pp. 87–112). Berlin: Mouton de Gruyter.

Research Article

Phonetica 2019;76:179–200
DOI: 10.1159/000494929

Received: April 3, 2017
Accepted: October 29, 2018
Published online: May 21, 2019

The Goldilocks Zone of Perceptual Learning

Molly Babel[a] Michael McAuliffe[b] Carolyn Norton[a] Brianne Senior[a]
Charlotte Vaughn[c]

[a]University of British Columbia, Vancouver, BC, Canada; [b]McGill University, Montreal, QC, Canada; [c]University of Oregon, Eugene, OR, USA

Abstract

Background/Aims: Lexically guided perceptual learning in speech is the updating of linguistic categories based on novel input disambiguated by the structure provided in a recognized lexical item. We test the range of variation that allows for perceptual learning by presenting listeners with items that vary from subtle within-category variation to fully remapped cross-category variation. ***Methods:*** Experiment 1 uses a lexically guided perceptual learning paradigm with words containing noncanonical /s/ realizations from s/ʃ continua that correspond to "typical," "ambiguous," "atypical," and "remapped" steps. Perceptual learning is tested in an s/ʃ categorization task. Experiment 2 addresses listener sensitivity to variation in the exposure items using AX discrimination tasks. ***Results:*** Listeners in experiment 1 showed perceptual learning with the maximally ambiguous tokens. Performance of listeners in experiment 2 suggests that tokens which showed the most perceptual learning were not perceptually salient on their own. ***Conclusion:*** These results demonstrate that perceptual learning is enhanced with maximally ambiguous stimuli. Excessively atypical pronunciations show attenuated perceptual learning, while typical pronunciations show no evidence for perceptual learning. AX discrimination illustrates that the maximally ambiguous stimuli are not perceptually unique. Together, these results suggest that perceptual learning relies on an interplay between confidence in phonetic and lexical predictions and category typicality.

Introduction

Lexically guided perceptual learning in speech uses novel input, often disambiguated by the linguistic scaffolding offered by a recoverable lexical frame, to update linguistic categories. It has been argued that perceptual learning is a means by which

KARGER

E-Mail karger@karger.com
www.karger.com/pho

Molly Babel
Department of Linguistics, University of British Columbia, 2613 West Mall, Totem Field Studios
Vancouver, BC V6T 1Z4 (Canada)
E-Mail molly.babel@ubc.ca

listeners manage the immense amounts of cross-talker variability in spoken language. When listeners are exposed to the novel input in sentences produced by nonnative accents, they rapidly learn and adapt in response to these nonnative accents, generalizing their implicitly updated phonetic and phonological knowledge to novel voices (Baese-Berk, Bradlow, & Wright, 2013; Bradlow & Bent, 2008; Clarke & Garrett, 2004). This adaptation occurs even for utterances that match on sublexical or syntactic levels (Cooper & Bradlow, 2016). Children also show the ability to rapidly adapt to unfamiliar accents (Holt & Bent, 2017), and both children and adults demonstrate better performance – in terms of speed or magnitude of learning – in sentences that are high in semantic predictability (Bradlow & Alexander, 2007; Holt & Bent, 2017).

Scaling downwards from the sentential level, listeners' enhanced ability to perceptually learn in linguistically predictable environments is also supported at the single-word level. Norris, McQueen, and Cutler (2003) introduced a paradigm for lexically guided perceptual learning in speech using individual lexical items. They synthesized a fricative sound that was ambiguous between /s/ or /f/ – denoted here as /?sf/ – and used it to replace word-final /s/ or /f/ fricatives in Dutch words. Listeners were then exposed to the ambiguous fricative in contexts where they would *expect* an /f/ or an /s/. They quantified their listeners' perceptual learning in a post-test categorization task with an /ɛs/–/ɛf/ continuum and demonstrated that listeners who were exposed to the ambiguous fricative in the context of /s/ words increased their perception of what was an acceptable /s/ in the post-test, while listeners exposed to the ambiguous /?sf/ in /f/ words likewise expanded their /f/ category to include these ambiguous sounds. Crucially, listeners exposed to the acoustically identical /?sf/ fricatives in nonwords showed no perceptual learning, as they had no linguistic structure to guide their interpretation of the ambiguous fricative. Note that in addition to lexical status assisting in the interpretation of a sound, the phonological position of the critical sound within the word also appears to be crucial. When ambiguous sounds replace critical fricatives in initial position, listeners do not show perceptual learning (Jesse & McQueen, 2011; McAuliffe & Babel, 2016). Similarly, when ambiguous sounds hamper a listener's ability to identify the item as a word (Clarke-Davidson, Luce, & Sawusch, 2008; Scharenborg & Janse, 2013; Witteman, Weber, & McQueen, 2013), the listener is less likely to perceptually learn. This lexically guided perceptual learning paradigm has been exploited considerably (Kraljic & Samuel, 2005, 2006, 2007; Maye, Aslin, & Tanenhaus, 2008; Reinisch, Weber, & Mitterer, 2013; Weatherholtz, 2015). Researchers have also demonstrated that visual information about place of articulation of a speech sound can disambiguate phonetically anomalous acoustic variation, supporting perceptual learning of speech sounds in the absence of lexical information (Vroomen, van Linden, Keetels, De Gelder, & Bertelson, 2004; Reinisch, Wozny, Mitterer, & Holt, 2014).

With this empirical background, we can consider the mechanisms proposed to account for lexically guided perceptual learning. A necessary component of a model of lexically guided perceptual learning is the association between the phonetic detail of an auditory object and its phonological category, determined in part by lexical context. In cases where we see perceptual learning, this phonetic detail must be used to update the multidimensional phonetic representation of a phonological category. The crux of the theoretical debate is how these connections are made. In TRACE, an interactive model of speech perception, McClelland and colleagues (McClelland & Elman, 1986; McClelland, Mirman, & Holt, 2006) argue that the activation of a lexical

was an English word or a nonword and to register their response via a button box. Button-label assignments were counterbalanced across participants (i.e., half the participants used the left button to respond with a word response and the other half the right button). If participants did not respond within the 3,000 ms time frame, a message appeared on screen informing them that no response had been detected, and the experiment progressed to the next trial. Listeners completed a total of 200 trials and were given self-paced breaks every 50 trials. Trial order was pseudorandomized to ensure the first 6 trials contained filler items and that critical or control items did not appear consecutively, with each listener in a condition receiving a different trial order. All listeners heard the same filler words, filler nonwords, and control /ʃ/ words in the exposure phase, but the critical /s/ items differed according to condition assignment; listeners were presented with either typical, ambiguous, atypical, or remapped critical /s/ items.

Test: Categorization Task. The test phase was structured similarly to the exposure phase. Listeners completed a two-alternative, forced-choice identification task. On each trial, they heard one of the 6 steps from one of the minimal word continua and were presented with two options on the screen, the "s" word (e.g., *sack*) or the corresponding "sh" word (e.g., *shack*). They were given 3,000 ms after stimulus onset to indicate their response by pressing the appropriate button on the response box. Participants completed 168 trials (4 continua × 6 steps × 7 repetitions of each step) and received self-paced breaks every 40 trials.

A control group of 25 participants completed just the test phase (from McAuliffe, 2015). Their results served as a baseline categorization function.

Experiment 1 Results

Lexical Endorsement in Exposure Phase

Responses faster than 200 ms or slower than 2,500 ms from the onset of the trial were excluded from analysis. Accuracy on the lexical decision task was high overall, 94% for filler words and 88% for nonwords. Trials with nonword stimuli were excluded for further analyses of accuracy. A logistic mixed effects model was used to analyze word recognition accuracy as the dependent variable, fitted in R (R Core Team, 2017) using the lme4 package (Bates, Maechler, Bolker, & Walker, 2015).[1] Fixed effects for exposure item type (filler, /s/, /ʃ/), and category typicality (typical, ambiguous, atypical, remapped) were used to assess the differences in word endorsement rates of the critical /s/ words compared to other item types across the different exposure conditions. Random effects for subjects and words, as well as random slopes for each were added where possible (by-speaker random slope for exposure item type and by-word random slope for category typicality). Both exposure item type and category typicality were contrast coded with filler and typical as the reference levels, respectively.

Within the typical category exposure, /s/ words were endorsed at a lower rate than filler items only marginally [B = –0.79; SE = 0.45; z = –1.76, p = 0.08]; however, there was significantly less endorsement of /s/ words compared to filler items for the ambiguous category [B = –1.24; SE = 0.60; z = –2.08; p = 0.04], atypical [B = –3.13; SE = 0.55; z = –5.66; p < 0.001], and remapped [B = –3.89; SE = 0.55; z = –7.04; p <

[1] Formula used: accuracy ~ exposure item type × category typicality + (1 + exposure item type | subject) + (1 + category typicality | word).

0.001]. A post hoc analysis of endorsement rates for /s/ words across conditions was done using the lsmeans package (Lenth, 2016), with *p* values adjusted using the Tukey method. Endorsement rates for /s/ words were not significantly different between typical and ambiguous [B = 0.37; SE = 0.48; z = 0.77; *p* = 0.89], but rates were significantly different between both ambiguous and atypical [B = 2.13; SE = 0.46; z = 4.66; *p* < 0.001], and atypical and remapped [B = 1.39; SE = 0.43; z = 3.21; *p* = 0.007]. Endorsement rates for filler words and control /ʃ/ words did not differ significantly across exposures [all *p* > 0.10]. These results are shown in Figure 2.

Categorization in Test Phase

Responses were coded as 1 for /s/ responses and 0 for /ʃ/ responses. Positive significant estimates therefore indicate a higher likelihood of /s/ response in categorization. Thus, positive significant effects are interpreted as indicative of perceptual learning, since a higher likelihood of /s/ response is associated with an expanded /s/ category. As above, responses faster than 200 ms or slower than 2,500 ms from the onset of the trial were excluded from analysis. Listeners' categorizations were analyzed in a logistic mixed effects model using step and exposure category typicality (control, typical, ambiguous, atypical, remapped) as predictors.[2] There were random intercepts for subject (with step as a random slope) and item (with step, typicality, and their interaction as random slopes). The control condition was used as the reference level. Step was centered, so effects around zero represent categorization behavior similar to the pretest participants (and the control participants), and effects above zero represent participants categorizing more of the continua as the /s/ word in the minimal pair rather than the /ʃ/ word.

The model had a nonsignificant intercept [B = 0.16; SE = 0.28; z = 0.58; *p* = 0.56] for the control condition. The ambiguous condition was significantly different from the control condition [B = 1.06; SE = 0.28; z = 3.67; *p* < 0.01], but the typical condition showed no effect [B = 0.15; SE = 0.30; z = 0.50; *p* = 0.62]. There were trending effects for both atypical [B = 0.48; SE = 0.30; z = 1.59; *p* = 0.11] and remapped [B = 0.48; SE = 0.29; z = 1.64; *p* = 0.10]. All interactions between step and category typicality were not significant [all *p* > 0.10].

Discussion

All effects trended towards perceptual learning (more of the continuum categorized as /s/), but the effect sizes in conditions other than ambiguous may be smaller than could be reliably found with the power in this study. The sample size used in this study (25 participants per condition) was based on previous work (McAuliffe & Babel, 2016; Reinisch et al., 2013) for reliably finding moderate effect sizes.

As shown in Figure 3a, the categorization function for the typical condition is the least different from the control condition, indicating a lack of perceptual learning for this group. Listeners presented with the atypical tokens and remapped tokens show moderate amounts of perceptual learning, but not significantly more than those in the control condition. Listeners in the ambiguous condition, however, show robust perceptual learning. Figure 3b plots the effect sizes of each typicality condition taken from the logistic mixed-effects model. These are the coefficients and standard errors

[2] Formula used: accuracy – step × category typicality+ (1 + step | subject) + (1 + step × category typicality | item).

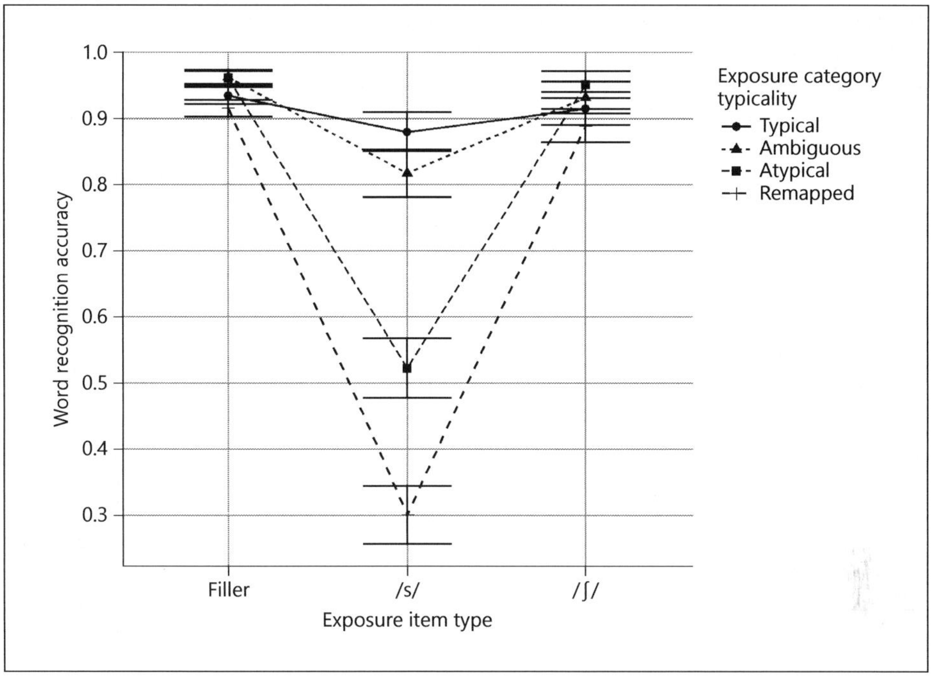

Fig. 2. Within-subject mean accuracy (*y* axis) for filler, /s/ words, and control /ʃ/ words (*x* axis) in the lexical decision exposure phase for the four experimental groups. Error bars represent 95% confidence intervals.

of the statistical models and are another visualization of the increased magnitude of perceptual learning for listeners exposed to items that together presented the maximally ambiguous /s/ distribution.

Listeners appear to learn the most from stimuli that are maximally ambiguous, where maximum ambiguity is determined by pretests in which listeners categorized word-nonword continua as words and nonwords. Does this mean that the maximally ambiguous stimuli are perceptually unique? In other words, are listeners maximally sensitive to this phonetic variant? To assess this, we conducted speeded AX discrimination tasks (i.e., where participants decide whether two items are same or different) with the whole-word stimulus listeners heard in the lexical decision exposure phase of the perceptual learning tasks (experiment 2A) and with the fricative-vowel sequences of those words excised as free-standing CV sequences (experiment 2B).

Experiments 2A and 2B: Categorical Perception

Experiment 2A: Categorical Perception with Words

Methods

Participants. Twenty listeners participated in this task. As in experiment 1, all participants were self-reported native English speakers recruited from the University

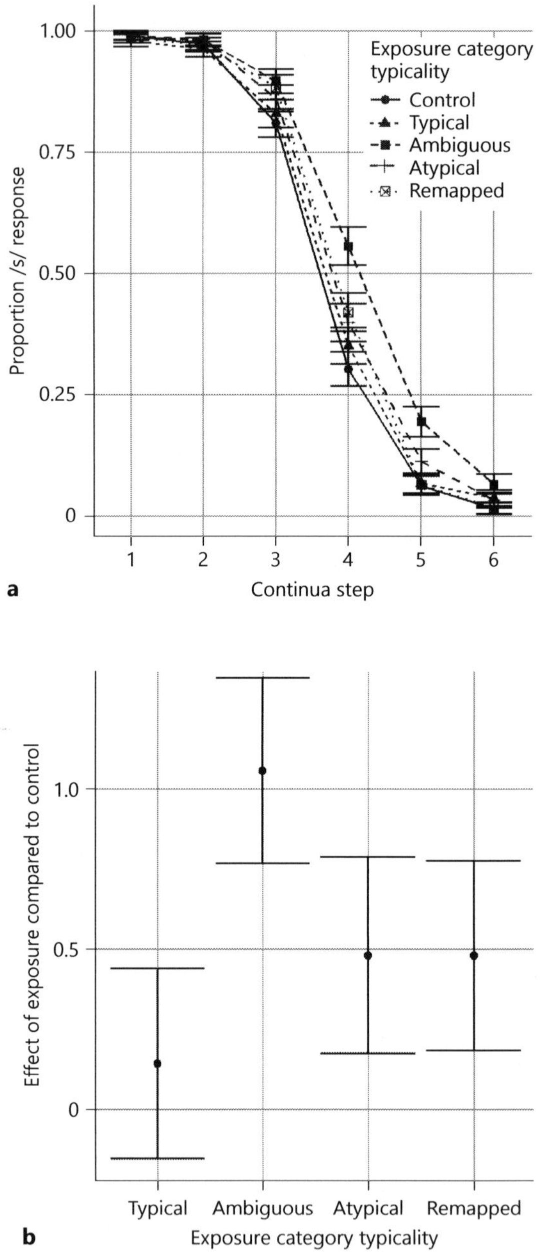

Fig. 3. a Proportion /s/ word responses (*y* axis) across the 6-step minimal pair continua (*x* axis) for the four experimental conditions and the control condition from McAuliffe (2015). Error bars represent 95% confidence intervals. **b** Effect sizes (*y* axis) across experimental conditions (*x* axis). Effect size is plotted as the coefficient and standard error from the mixed effects models. Experimental condition is ordered from most typical (typical) to least typical (remapped) /s/.

of British Columbia community and reported no known speech, language, or hearing impairments. Participants were compensated with partial course credit for their participation. None of the listeners in this experiment had participated in experiment 1.

Stimuli. Stimuli for the whole-word categorization task consisted of 40 critical words containing a word-initial (e.g., *celery*) or word-medial /s/ (e.g., *cursor*, from McAuliffe & Babel, 2016) that were used in the exposure phase of experiment 1. Stimuli maintained the category typicality steps corresponding to typical, ambiguous, atypical, and remapped categories used in experiment 1, along with a new set of "canonical" items (100% /s/ word endorsement from the pretest).

Procedure. Participants were seated up to four at a time at individual PC workstations outfitted with AKG-K240 headphones and a PST serial response box. Upon completion of the perception task, participants completed a language background questionnaire.

Participants received instructions orally and visually on the desktop prior to the experiment, instructing them to complete an AX categorization task of whole-word pairs. Stimuli contained a critical word-medial or word-initial /?sʃ/ fricative. Listeners were presented with pairs of adjacent different stimuli (e.g., canonical/typical, typical/ambiguous, ambiguous/atypical, atypical/remapped; "different trials") or pairs of identical stimuli ("same trials") with a 500-ms interstimulus interval. Participants were given up to 5,000 ms to indicate whether the auditory stimulus pairs were the "same" or "different" by pressing the appropriate button on the response box (e.g., 1 = same, 5 = different). Button-label assignments were counterbalanced across participants. Participants received visual feedback (accuracy and response time) following each trial. After every 50 trials, participants were provided with a self-administered break. Participants completed a total of 544 trials (272 same; 272 different). Not all words had the same number of comparisons because of the shape of the word endorsement response functions for some items. As reported in Table 1 in the Appendix, the continuum steps associated with adjacent categories were identical for some items (i.e., the response function was flat across a range of steps).

Experiment 2A Results

Although listeners responded to stimuli containing /s/ in both initial (e.g., *silver*, /sɪlvɹ̩/) and medial (e.g., *croissant*, /kɹəsɑnt/) positions, we restricted this analysis to responses to /s/-medial stimuli in order to be maximally comparable to the exposure stimuli in experiment 1. Response time outliers, defined as responses more than 2 standard deviations away from the group mean, were removed, eliminating 4.8% of the data. The remaining trials have a mean response time of 1,064 ms (SD = 287). Accuracy on the remaining trials was used to calculate the signal detection theoretic values d' and c for the analysis of the categorical perception data. d' is a measure of sensitivity to the signal versus noise based on the inverse of the cumulative distribution function of the hit rate (H) and false alarm rate (FA), with higher values of d' indicating greater sensitivity. It was calculated for roving same-different discrimination tasks following Macmillan and Creelman (2004) using the psyphy package in R (Knoblauch, 2014). c, a measure of response bias, was calculated as $-0.5 \times (z(H) + z(FA))$, where z refers to the inverse of the cumulative distribution function; based on how we calculated these values, a positive value indicates a bias to respond "same."

For the sensitivity analysis, d' values from the word and CV experiments were analyzed in a repeated-measures ANOVA with pair (e.g., canonical/typical, typical/

Phonetica 2019;76:179–200
DOI: 10.1159/000494929

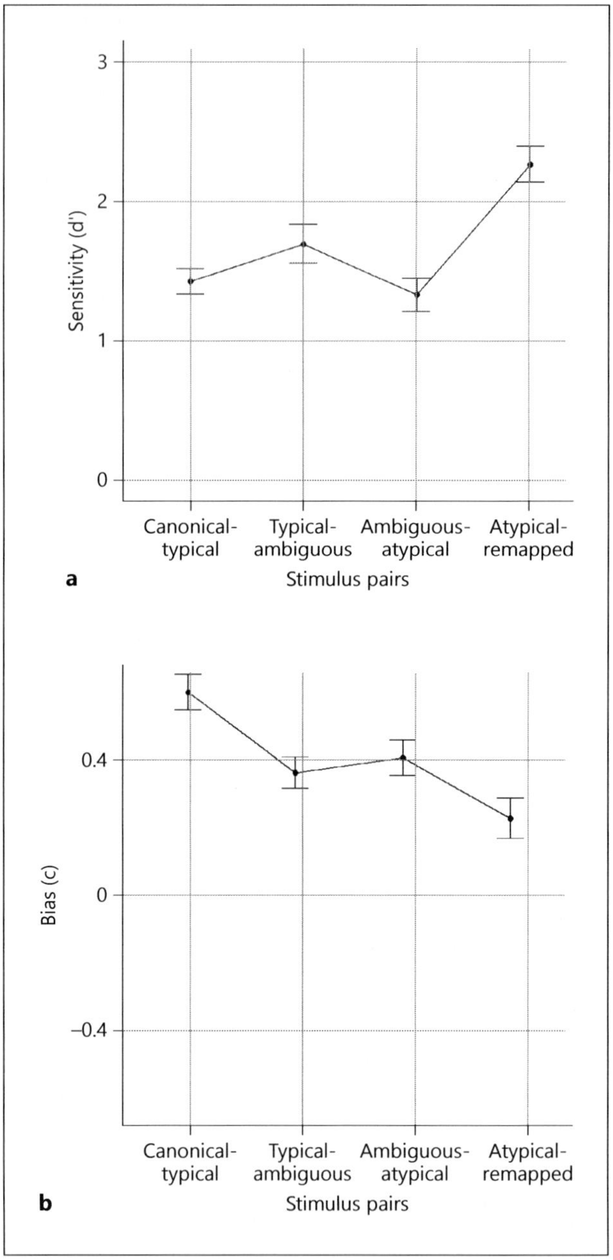

Fig. 4. Experiment 2A (words). **a** d', a measure of sensitivity, on the *y* axis for each stimulus pair type (*x* axis). Higher values indicate higher perceptual sensitivity. **b** c, a measure of response bias, on the *y* axis for stimulus pairs (*x* axis). Error bars show standard errors.

ambiguous, etc.) repeated across listeners. The d' analysis showed a main effect of pair [$F(3, 99) = 21.24$, $p < 0.001$]. The Tukey honest significant difference test was used to compare sensitivity between stimulus pairs. Listeners discriminated the atypical/remapped pair more accurately than canonical/typical ($p < 0.001$), typical/ambiguous

DOI: 10.1159/000494929

($p = 0.006$), and ambiguous/atypical ($p < 0.001$) pairs. None of the other comparisons within each stimulus type were significantly different in the Tukey tests. These results, shown in Figure 4a, indicate that listeners were most accurate at discriminating between pairs on the /ʃ/ side of the continuum.

To examine listeners' response bias across the continuum pairs, we computed and analyzed c with pair as a within-listener variable. The analysis for c revealed a main effect of pair [$F(3, 99) = 18.23$, $p < 0.001$]. As shown in Figure 4b, listeners showed a more numerically gradual change in bias across the pairs. Post hoc Tukey tests indicate that listeners were more biased to respond "same" on the far word side of the continuum. The greatest bias towards a "same" response occurred with canonical/typical pairs, as compared to typical/ambiguous ($p = 0.01$), ambiguous/atypical ($p = 0.05$), and atypical/remapped ($p < 0.001$) pairs. All other comparisons for these stimuli were not significant.

Experiment 2B: Categorical Perception with CVs

The results from experiment 2A provide no support for the maximally ambiguous step being perceptually unique to listeners. However, it is possible that listeners' lexical and linguistic biases clouded their ability to perceive small differences between different items (Ganong, 1980; Pisoni & Tash, 1974). To address this we ran a nearly identical experiment, presenting listeners with CV sequences composed of the /?s/ fricatives and the immediately following vocalic unit to determine whether the maximally ambiguous step stood out as perceptually unique in the absence of lexical bias.[3]

Methods

Participants. Seventeen novel listeners who had not previously participated in either of the previous experiments completed this task. All participants were self-reported native English speakers with no known speech, language, or hearing impairments, and they were recruited from the same population as in experiments 1 and 2A. Participants were compensated with partial course credit for their participation.

Stimuli. The stimuli consisted of 40 CV sequences excised from the 40 words used in experiment 2A. CV sequences were excised using the following boundaries: the onset of the fricative was identified as the onset of aperiodic energy, and the offset of the following vocoid was identified as the cessation of periodic energy. Of the 544 trials presented to participants, 288 trials were of CV sequences excised from fricative medial words.

Procedure. The procedure was identical to that of experiment 2A.

Experiment 2B Results

Analyses were conducted on the CV sequences excised from the fricative medial stimuli. Responses which made more than 2 SDs from the overall mean response time

3 A reviewer points out that many of these excised CV sequences themselves create words (e.g., /si/ and /ʃi/ excised from the canonical and remapped end points of galaxy create the words see and she). Including fricative-rhotic sequences, which create words like sure and sir, 9 of the 20 CV sequences mapped to real CV words. Being excised out of word-medial positions from multisyllabic words, these items retain any coarticulation from the surrounding context and are considerably shorter in duration (mean: 275 ms, SD: 54 ms) than the real word equivalents of these words (for example, test items sigh and shy were 635 and 622 ms, respectively). This fact and the monotony of the experiment make it less likely that listeners processed these items as fully word-like (Cutler et al., 1987).

Phonetica 2019;76:179–200
DOI: 10.1159/000494929

were removed, constituting 5.2% of the data. The remaining trials had a mean response time of 866 ms (SD = 317). Then d' and c were calculated on these trials. For the sensitivity analysis, d' values were analyzed in a repeated measures ANOVA with pair repeated across listeners. The analysis showed a main effect of pair [$F(3, 78) = 35.25, p < 0.001$]. The Tukey HSD test was used to compare sensitivity between stimulus pairs. Like in experiment 2A, listeners discriminated the atypical/remapped pairs more accurately than: canonical/typical ($p = 0.001$), typical/ambiguous ($p < 0.001$), and ambiguous/atypical ($p < 0.001$) pairs. None of the other comparisons within each stimulus type were significantly different from each other. Like with the whole-word stimuli, listeners were most accurate on the /ʃ/ side of the continuum, as shown in Figure 5a.

Listeners' response biases were analyzed with c as the dependent measure and stimulus pair as a within-listener independent variable. This analysis returned a main effect of pair [$F(3, 78) = 15.97, p < 0.001$]. As can be seen in Figure 5b, listeners were least biased towards a same response with the atypical/remapped pair compared to ambiguous/atypical ($p = 0.01$), typical/ambiguous ($p = 0.004$), and canonical/typical ($p = 0.01$). None of the other comparisons were significant.

To assess whether sensitivity or bias were related to response time, we compared a listener's mean response time for each pair to their d' and c values. There was a tendency for slower response times to be associated with greater perceptual sensitivity [$r(106) = 0.19, p = 0.05$], and there was no relationship between bias and response time [$r(106) = -0.03, p = 0.72$].

Discussion of Experiments 2A and 2B

To assess whether the maximally ambiguous items from experiment 1 were perceptually distinct we conducted two AX discrimination tasks with the whole words (experiment 2A) and excised fricative vowel sequences (experiment 2B) from items with canonical, typical, ambiguous, atypical, or remapped endorsement rates. Perceptual sensitivity was quantified using d', and response bias was evaluated using c.

Listeners' perceptual sensitivity with these stimuli peaked with the atypical/remapped pair, indicating that it was not until the fricative was fully remapped to /ʃ/ that listeners showed a clear perceptual boundary. While sensitivity was unaffected by stimulus type, listener response bias was affected. An interaction between stimulus type and pair indicated that when presented with stimuli as whole words, listeners were maximally biased to respond "same" for the canonical/typical pairs, indicating a "same" bias on the side of the continuum that is more word-like. Listeners who did the discrimination task with CVs excised from the word items showed the opposite pattern; when presented with excised CV sequences, listeners' bias to respond "same" remained stable across the first three continuum pairs and decreased with the atypical/remapped pair on the /ʃ/ side of the continua. Together, these results indicate that the maximally ambiguous fricatives are not perceptually unique in either a sensitivity or bias analysis. While not a corroboration of these results, note that, as reported in Table 1 in the Appendix, the average step selected for the ambiguous items is closer to the average step for the typical and atypical items than the typical and atypical items are to the canonical and remapped items, respectively. This indicates that the end points of the continua were more perceptually unique in the categorization pretest as well.

Appendix

Table 1

Language Background Questionnaire Please answer the questions below to the best of your ability. Please ask the experimenter if you have any questions or concerns.	
1.	Are you a native speaker of English? In this case, "native" refers to your first language. Yes/No
2.	If English is not your native language, is it your dominant language? Yes/No
3.	If English is not your native language, what is/are your native language(s)?
4.	Regardless of whether English is your native or dominant language, what variety of English do you speak? Please specify a dialect (e.g., Newfoundland English, Southern US English, etc.) if you would like. ___American English ___Australian English ___British English ___Canadian English ___Indian English ___Irish English ___Hong Kong English ___Jamaican English ___New Zealand English ___Scottish English ___Singaporean English ___South African English ___Other. Please, specify:
5.	What gender do you identify as? ______________
6.	What is your racial or ethnic heritage? Check all that apply ___First Nations ___Asian ___Pacific Islander ___Black ___White ___Hispanic ___South Asian ___Other. Please, specify: ________
7.	What is your age? _____
8.	Are you right-handed or left-handed? ____
9.	What cities or towns have you lived in? Beginning with the place where you were born, please list each town or city (and country, if appropriate) you have lived in for 6 months or more.

10.	What is your proficiency in English? (1) not at all, (2) poorly, (3) fairly well, (4) fluently. Reading ___ Writing ____ Speaking ___ Listening ___
11.	At what age did you start learning English? _____
12.	Do you have knowledge of any languages other than English? This can include both languages you speak natively and ones you have learned in educational settings. Yes/No
13.	What other languages do you have knowledge of? Please include both languages you speak natively and ones you have learned in educational settings. When did you start learning this language? How would you rate your proficiency in reading, writing, speaking, and understanding it? (1) Not at all, (2) poorly, (3) fairly well, (4) fluently
14.	Which language(s) do you most commonly speak: At home? At work? At school? With friends? With parents? With grandparents?
15.	Do you have any speech or hearing disorders? If "yes", please specify:
16.	Where were your caretakers born and raised?
17.	What are your caretakers' first languages?
18.	What is the highest educational degree you have earned (or are in the process of earning)?
19.	What did you think the experiment was about? (Optional)

References

Allen, J. S., & Miller, J. L. (2004). Listener sensitivity to individual talker differences in voice-onset-time. *The Journal of the Acoustical Society of America, 115*(6), 3171–3183.

Baese-Berk, M. M., Bradlow, A. R., & Wright, B. A. (2013). Accent-independent adaptation to foreign accented speech. *The Journal of the Acoustical Society of America, 133*(3), EL174–EL180. doi: 10.1121/1.4789864

Bates, D., Maechler, M., Bolker, B., & Walker, S. (2015). Fitting linear mixed-effects models using lme4. *Journal of Statistical Software, 67*(1), 1–48. doi:10.18637/jss.v067.i01

Bradlow, A. R., & Alexander, J. A. (2007). Semantic and phonetic enhancements for speech-in-noise recognition by native and non-native listeners. *The Journal of the Acoustical Society of America, 121*(4), 2339–2349. doi:10.1121/1.2642103

Bradlow, A. R., & Bent, T. (2008). Perceptual adaptation to non-native speech. *Cognition, 106*(2), 707–729. doi:10.1016/j.cognition.2007.04.005

Clark, A. (2013). Whatever next? Predictive brains, situated agents, and the future of cognitive science. *Behavioral and Brain Sciences, 36*(3), 181–204. doi:10.1017/S0140525X12000477

Clarke, C. M., & Garrett, M. F. (2004). Rapid adaptation to foreign-accented English. *The Journal of the Acoustical Society of America, 116*(6), 3647–3658. doi:10.1121/1.1815131

Clarke-Davidson, C. M., Luce, P. A., & Sawusch, J. R. (2008). Does perceptual learning in speech reflect changes in phonetic category representation or decision bias? *Attention, Perception & Psychophysics, 70*(4), 604–618. doi:10.3758/PP.70.4.604

Cooper, A., & Bradlow, A. R. (2016). Linguistically guided adaptation to foreign-accented speech. *The Journal of the Acoustical Society of America, 140*(5), EL378–EL384. doi:10.1121/1.4966585

Cutler, A. (2012). *Native Listening: Language Experience and the Recognition of Spoken Words*. Cambridge: MIT Press.

Cutler, A., Mehler, J., Norris, D., & Segui, J. (1987). Phoneme identification and the lexicon. *Cognitive Psychology, 19*(2), 141–177. doi:10.1016/0010-0285(87)90010-7

Ganong, W. F., III. (1980). Phonetic categorization in auditory word perception. *Journal of Experimental Psychology. Human Perception and Performance, 6*(1), 110–125. doi:10.1037/0096-1523.6.1.110

Hay, J. B., Pierrehumbert, J. B., Walker, A. J., & LaShell, P. (2015). Tracking word frequency effects through 130 years of sound change. *Cognition, 139*, 83–91. doi:10.1016/j.cognition.2015.02.012

Holt, R. F., & Bent, T. (2017). Children's use of semantic context in perception of foreign-accented speech. *Journal of Speech, Language, and Hearing Research: JSLHR, 60*(1), 223–230. doi:10.1044/2016_JSLHR-H-16-0014

Jesse, A., & McQueen, J. M. (2011). Positional effects in the lexical retuning of speech perception. *Psychonomic Bulletin & Review, 18*(5), 943–950. doi:10.3758/s13423-011-0129-2

Kawahara, H, Morise, M, Takahashi, T, Nisimura, R, Irino, T, & Banno, H (2008, March). TANDEM-STRAIGHT: A temporally stable power spectral representation for periodic signals and applications to interference-free spectrum, F0, and aperiodicity estimation. *ICASSP. 2008 IEEE International Conference on Acoustics, Speech and Signal Processing*; 2008 March 31 to April 4; Las Vegas (NE), pp. 3933-3936.

Kleinschmidt, D. F., & Jaeger, T. F. (2015). Robust speech perception: Recognize the familiar, generalize to the similar, and adapt to the novel. *Psychological Review, 122*(2), 148–203. doi:10.1037/a0038695

Knoblauch, K. (2014). psyphy: Functions for analyzing psychophysical data in R. R package version 0.1-9. Retrieved from https://CRAN.R-project.org/package=psyphy

Kraljic, T., & Samuel, A. G. (2005). Perceptual learning for speech: Is there a return to normal? *Cognitive Psychology, 51*(2), 141–178. doi:10.1016/j.cogpsych.2005.05.001

Kraljic, T., & Samuel, A. G. (2006). Generalization in perceptual learning for speech. *Psychonomic Bulletin & Review, 13*(2), 262–268. doi:10.3758/BF03193841

Kraljic, T., & Samuel, A. G. (2007). Perceptual adjustments to multiple speakers. *Journal of Memory and Language, 56*(1), 1–15. doi.org/10.1016/j.jml.2006.07.010

Kraljic, T., Brennan, S. E., & Samuel, A. G. (2008a). Accommodating variation: Dialects, idiolects, and speech processing. *Cognition, 107*(1), 54–81. doi:10.1016/j.cognition.2007.07.013

Kraljic, T., Samuel, A. G., & Brennan, S. E. (2008b). First impressions and last resorts: How listeners adjust to speaker variability. *Psychological Science, 19*(4), 332–338. doi:10.1111/j.1467-9280.2008.02090.x

Kuperberg, G. R., & Jaeger, T. F. (2016). What do we mean by prediction in language comprehension? *Language, Cognition and Neuroscience, 31*(1), 32–59. doi:10.1080/23273798.2015.1102299

Lenth, R. (2016). Least-squares means: The R package lsmeans. *Journal of Statistical Software, 69*(1), 1–33. doi:10.18637/jss.v069.i01

Liberman, A. M., Harris, K. S., Hoffman, H. S., & Griffith, B. C. (1957). The discrimination of speech sounds within and across phoneme boundaries. *Journal of Experimental Psychology, 54*(5), 358–368. doi:10.1037/h0044417

Macmillan, N. A., & Creelman, C. D. (2004). *Detection theory: A user's guide*. New York: Psychology Press.

Maye, J., Aslin, R. N., & Tanenhaus, M. K. (2008). The weckud wetch of the wast: Lexical adaptation to a novel accent. *Cognitive Science, 32*(3), 543–562. doi:10.1080/03640210802035357

McAuliffe, M. (2015). *Attention and salience in lexically guided perceptual learning* (University of British Columbia doctoral dissertation). cIRcle: UBC's Digital Repository: Electronic Theses and Dissertations (ETDs) 2008+. Retrieved from: http://hdl.handle.net/2429/54152

McAuliffe, M., & Babel, M. (2016). Stimulus-directed attention attenuates lexically-guided perceptual learning. *The Journal of the Acoustical Society of America, 140*(3), 1727–1738. doi:10.1121/1.4962529

McClelland, J. L., & Elman, J. L. (1986). The TRACE model of speech perception. *Cognitive Psychology, 18*(1), 1–86. doi:10.1016/0010-0285(86)90015-0

McClelland, J. L., Mirman, D., & Holt, L. L. (2006). Are there interactive processes in speech perception? *Trends in Cognitive Sciences, 10*(8), 363–369. do:10.1016/j.tics.2006.06.007

Norris, D., McQueen, J. M., & Cutler, A. (2000). Merging information in speech recognition: Feedback is never necessary. *Behavioral and Brain Sciences, 23*(3), 299–325. doi:10.1017/S0140525X00003241

Norris, D., McQueen, J. M., & Cutler, A. (2003). Perceptual learning in speech. *Cognitive Psychology, 47*(2), 204–238. doi:10.1016/S0010-0285(03)00006-9

Norris, D., McQueen, J. M., & Cutler, A. (2016). Prediction, Bayesian inference and feedback in speech recognition. *Language, Cognition and Neuroscience, 31*(1), 4–18. doi:10.1080/23273798.2015.1081703

Pisoni, D. B., & Tash, J. (1974). Reaction times to comparisons within and across phonetic categories. *Attention, Perception & Psychophysics, 15*(2), 285–290. doi:10.3758/BF03213946

Pitt, M. A., & Samuel, A. G. (1993). An empirical and meta-analytic evaluation of the phoneme identification task. *Journal of Experimental Psychology. Human Perception and Performance, 19*(4), 699–725. doi:10.1037/0096-1523.19.4.699

Psychology Software Tools Inc. (2012). *E-Prime 2.0* [computer software]. Retrieved from http://www.pstnet.com

R Core Team (2017). *R: A language and environment for statistical computing.* R Foundation for Statistical Computing, Vienna, Austria. Retrieved from https://www.R-project.org/

Reinisch, E., Weber, A., & Mitterer, H. (2013). Listeners retune phoneme categories across languages. *Journal of Experimental Psychology. Human Perception and Performance, 39*(1), 75–86. doi:10.1037/a0027979

Reinisch, E., Wozny, D. R., Mitterer, H., & Holt, L. L. (2014). Phonetic category recalibration: What are the categories? *Journal of Phonetics, 45*, 91–105. doi:10.1016/j.wocn.2014.04.002

Scharenborg, O., & Janse, E. (2013). Comparing lexically guided perceptual learning in younger and older listeners. *Attention, Perception & Psychophysics, 75*(3), 525–536. doi:10.3758/s13414-013-0422-4

Scharenborg, O., Weber, A., & Janse, E. (2015). The role of attentional abilities in lexically guided perceptual learning by older listeners. *Attention, Perception & Psychophysics, 77*(2), 493–507. doi:10.3758/s13414-014-0792-2

Sohoglu, E., & Davis, M. H. (2016). Perceptual learning of degraded speech by minimizing prediction error. *Proceedings of the National Academy of Sciences of the United States of America, 113*(12), E1747–E1756. doi:10.1073/pnas.1523266113

Sóskuthy, M. (2015). Understanding change through stability: A computational study of sound change actuation. *Lingua, 163*, 40–60. doi:10.1016/j.lingua.2015.05.010

Sumner, M. (2011). The role of variation in the perception of accented speech. *Cognition, 119*(1), 131–136. doi.org/10.1016/j.cognition.2010.10.018

Theodore, R. M., Myers, E. B., & Lomibao, J. A. (2015). Talker-specific influences on phonetic category structure. *The Journal of the Acoustical Society of America, 138*(2), 1068–1078. doi:10.1121/1.4927489

Vaughn, C., & Kendall, T. (2018). Listener sensitivity to probabilistic conditioning of sociolinguistic variables: The case of (ING). *Journal of Memory and Language, 103*, 58–73. doi.org/10.1016/j.jml.2018.07.006

Vroomen, J., van Linden, S., Keetels, M., De Gelder, B., & Bertelson, P. (2004). Selective adaptation and recalibration of auditory speech by lipread information: Dissipation. *Speech Communication, 44*(1), 55–61. doi.org/10.1016/j.specom.2004.03.009

Warren, R. M. (1970). Perceptual restoration of missing speech sounds. *Science, 167*(3917), 392–393. doi.org/10.1126/science.167.3917.392

Weatherholtz, K. (2015). *Perceptual learning of systemic cross-category vowel variation* [Unpublished doctoral dissertation]. Columbus, OH: The Ohio State University.

Wedel, A. (2012). Lexical contrast maintenance and the organization of sublexical contrast systems. *Language and Cognition, 4*(4), 319–355. doi:10.1515/langcog-2012-0018

Witteman, M. J., Weber, A., & McQueen, J. M. (2013). Foreign accent strength and listener familiarity with an accent codetermine speed of perceptual adaptation. *Attention, Perception & Psychophysics, 75*(3), 537–556. doi:10.3758/s13414-012-0404-y

Xie, X., Theodore, R. M., & Myers, E. B. (2017). More than a boundary shift: Perceptual adaptation to foreign-accented speech reshapes the internal structure of phonetic categories. *Journal of Experimental Psychology. Human Perception and Performance, 43*(1), 206–217. doi:10.1037/xhp0000285

Research Article

Phonetica 2019;76:201–233
DOI: 10.1159/000494927

Received: March 28, 2017
Accepted: October 29, 2018
Published online: May 21, 2019

Phonetic Realisation and Phonemic Categorisation of the Final Reduced Corner Vowels in the Finnic Languages of Ingria

Natalia Kuznetsova[a, b] Vasilisa Verkhodanova[c]

[a]Institute for Linguistic Studies, Department of the Languages of Russia, Russian Academy of Sciences, St. Petersburg, Russia; [b]Dipartimento di Lingue e Letterature straniere e Culture moderne, Universita degli Studi di Torino, Turin, Italy; [c]Rijksuniversiteit Groningen, Campus Fryslân, Leeuwarden, The Netherlands

Abstract

Individual variability in sound change was explored at three stages of final vowel reduction and loss in the endangered Finnic varieties of Ingria (subdialects of Ingrian, Votic and Ingrian Finnish). The correlation between the realisation of reduced vowels and their phonemic categorisation by speakers was studied. The correlated results showed that if V was pronounced >70%, its starting loss was not yet perceived, apart from certain frequent elements, but after >70% loss, V was not perceived any more. A split of 50/50 between V and loss in production correlated with the same split in categorisation. At the beginning of a sound change, production is, therefore, more innovative, but after reanalysis, categorisation becomes more innovative and leads the change. The vowel *a* was the most innovative in terms of loss, *u*/*o* were the most conservative, and *i* was in the middle, while consonantal palatalisation was more salient than labialisation. These differences are based on acoustics, articulation and perception.

1. Introduction

1.1 Synchronic Variation and Sound Change

Systematic individual variation in speech perception and production produces a pool of variation which becomes the source of language change (Kruszewski, 1883; Baudouin de Courtenay, 1895; Ohala, 1989; Labov, 1999; Baker et

Natalia Kuznetsova
Dipartimento di Lingue e Letterature straniere e Culture moderne, Universita degli Studi di Torino
via Giuseppe Verdi, 10, IT–10124 Turin (Italy)
E-Mail nkuzn@yandex.ru

al., 2011; Yu, 2013; Stevens & Harrington, 2014; Bybee, 2015). Language change is propagated through the repeated exposure of several generations of speakers to a gradually changing variable pool of realisations. Learning theories, placed on a continuum between rational Bayesian approaches and associationist models inspired by biological discoveries, are now at the core of psychophysical sound change models.

Associative learning in phonology implies constant bidirectional updating of the connection weights in mappings between acoustic cues and phonological/subphonemic categories. Learning is distributional in that the learner acquires knowledge of the frequency distribution of various phonetic stimuli and builds a mental phonological model of the language on this. Frequency distribution is even suggested to be a more important factor in the formation of phonemic categories than minimal pairs (Maye & Gerken, 2000; Vallabha et al., 2007; Wanrooij et al., 2013; Olejarczuk et al., 2018). This approach also explains the puzzling cases of near-mergers, when speakers already categorise items into the same phonemic class when there is still a phonetic difference in the realisation of two former classes (Labov et al., 1991; Barnes, 2006; Roettger et al., 2014). The exact structure of such mental constructions is still, however, under debate between prototype, exemplar and other models (Mompeán-Gonzalez, 2004; Gureckis & Goldstone, 2008; Johnson, 2015; Davis & Poldrack, 2014; Kapatsinski, 2018).

Sound change, as any language change, follows the S-curve path, where a weighting jumps to a different value at some point during the change (Hyman, 1976; Kirby, 2010, p. 148; Blythe & Croft, 2012, p. 293). Its actuation is discussed (Baker et al., 2011; Stevens & Harrington, 2014; Priva, 2017), as well as the exact mechanism of the jump between values. The latter might be linked to properties of the articulatory/acoustic relation, when the manipulation of an articulator can result in non-monotonic variability in an acoustic parameter (Stevens, 2004). The usage-based approaches have also followed the hypothesis that lexical frequency prompts sound change. The latter starts from frequent words and morphemes due to a higher level of automatisation in their production and can later spread throughout the entire phonological system (van Bergem, 1995; Bybee, 2001, pp. 11–12; Bybee et al., 2016; Hay & Foulkes, 2016; Hall et al., 2018; Kapatsinski, 2018).

Sound change implies two connected processes: (1) a change in the structure of the pool of phonetic realisations and (2) a categorical reanalysis in the mind of the speakers/listeners. The temporal and causal correlations between these two processes are still unclear (Bybee, 2001, p. 55). Modern phonology still has to reconcile the data on the continuous and variable nature of the phonetic signal and on the behaviour of symbolic processes in a consistent fashion (Barnes, 2006, p. 222; Kirby, 2010, p. 149). In the associative learning framework, the same question concerns the relations between typicality distributions in perception and frequency distributions in production (Kapatsinski, 2018, p. 275). The concept of attractor landscape used in non-linear dynamic systems might be of use in modelling this link between continuous and categorical variation. A dynamic system is continuous, but there are specific stable states (attractors) it moves to (Ritter et al., 2018). A change in the weighting of attractors can model the change in the frequency distributions of different realisations throughout the sound change.

The present paper explores the correlation between production and mental representation in a case study on vowel reduction and loss in several minor Finnic varieties.

1.2 Vowel Reduction: General and Particular Mechanisms

Vowel reduction and loss is a cross-linguistically frequent phenomenon. However, studies taking a cross-linguistic and general theoretical approach to it are still scarce. Lindblom (1963) suggested that vowel reduction occurs through the mechanism of formant undershoot, which is a function of decrease in vowel duration. This view was supported by Delattre (1969), Flemming (1995, 2004), Kirchner (1998) and Barnes (2006), although the causal relation between undershoot and duration was reversed by Crosswhite (2004). The matter of reduction is discussed in a number of functionalist works, where the language system is represented as a trade-off between the needs of the speaker to economise the effort and the listener to be able to decipher the message. Lindblom (1990), for example, proposed an H&H framework, where a message varies in articulatory clarity as a compromise between hypospeech, minimising articulatory effort, and hyperspeech, maximising discriminability. Reduction characterises pieces of speech with low informativity and is a manifestation of hypospeech, a "part of planned speech behaviour rather than an accidental by-product of vocal organ inertia" (Harris, 2005, p. 132).

Reduction does not affect all vowel qualities or positions in a word or phrase equally, nor does it work always in the same direction. For example, a word-final and especially a phrase-final position manifests both vowel strengthening (lengthening and strengthening of articulation) and vowel weakening (devoicing, laryngealisation, nasalisation and loss). Barnes (2006) explains the weakening effects by the perceptual weakness of final vowels, in spite of their possible articulatory strength. Vowel reduction could also have different underlying mechanisms. Kapatsinski (2018, p. 286) opposes phonetically gradual reduction produced by automatisation of execution in production to phonetically abrupt loss of low-salience parts left meaningless by overshadowing in perception.

Two general paths of vowel reduction are distinguished: centripetal (centralisation towards schwa) and centrifugal (dispersion towards the three corner vowels *a*, *i*, and *u*). The corner vowels are generally known to be special in various respects: the most stable and focalised, perceptually salient, the easiest for neural processing because of maximal distinction, etc. (Crosswhite, 2004; Polka & Bohn, 2003, 2011; Harris, 2005; Johnson, 2015; Manca & Grimaldi, 2016). Data on acoustic, perceptual and other differences within the corner vowels are, however, scarce. The typological studies on vowel reduction show that vowel height is affected before frontness/backness, rounding or advanced tongue root contrasts (Barnes, 2006; Flemming, 2004). Reduced speech is characterised by compression of the acoustic space between F1 and F2 through F1 raising, an effect of lessened jaw opening (Lindblom, 1963; Uchanski, 2005). The bottom-up direction of the compression suggests that high unstressed vowels would be less marked than non-high ones (Walker, 2011, p. 29). The latter require more jaw opening and a longer time to be realised. At the same time, reduction-based sonority scales presume that the vowel *a* is less marked, but that schwa is more marked than *i* and *u* (Crosswhite, 2004, p. 209; de Lacy, 2006, p. 286).

The existence of differences between *i* and *u* is not much discussed in the surveys on vowel reduction. Some argue for a role of F2-based harmony in blocking the reduction of front vowels (Pearce, 2008; Szeredi, 2010). Evidence for the disparity between *i* and *u* comes also from research on vowel perception and neuroimaging, where place of articulation and tongue height are seen as relatively simple features. They directly correspond to F1 and F2 values, which, in turn, find their straight correlates in regions and types of brain activity. The rounding feature appears more complex, as it requires higher-level information processing and is acoustically less reliable and perceived with significant help from the visual channel (Traunmüller & Öhrström, 2007; Eulitz & Obleser, 2007; Vatakis et al., 2012; Manca & Grimaldi, 2016). One might thus suggest that *u* is less perceptually robust and salient than *i* and, therefore, more prompt for reduction, especially in languages with fronting vowel harmony.

Our study offers further experimental data to explore the general mechanisms of reduction and loss, as well as vowel markedness hierarchies at different subsequent stages of reduction.

2. Methods

2.1 Aims, Data and Background of the Study

Correlations between the frequencies of various realisations of the three corner vowels in production and mental categorisation were explored in a comparative phonetic field study (2014–2016) on final vowel reduction and loss. We looked at three Finnic languages of the Lower Luga area in the west of historical Ingria (currently, the vicinity of St. Petersburg in Russia): Ingrian, Votic and Finnish (see maps in Kuznetsova et al., 2015, p. 128; Kuznetsova & Sidorkevič, 2012, p. 565). They have been in close contact for centuries and formed a Lower Luga Sprachbund (Muslimov, 2005). Besides, a group of Ingrian and Finnish speakers was expelled from this region to Western Siberia in 1803–1804 after a strike against Baron von Ungern-Sternberg. A contact Siberian Ingrian/Finnish language developed there in isolation from its sister varieties (Nirvi, 1972; Sidorkevič, 2013).

The process of reduction advances through several stages, still observed in the living varieties of these languages. The following varieties were chosen for this study: (1) the Kurkola Ingrian Finnish dialect (IF); (2) the Luutsa dialect of Votic (V); (3) the Central (CI) and (4) the Southern (SI1, SI2) varieties of the Lower Luga dialect of Ingrian; and (5) Siberian Ingrian/Finnish (S). The data were obtained from 1 speaker per variety, with the exception of South Lower Luga Ingrian, for which 2 speakers were recorded (Table 1).

This is a limitation of our study, precipitated by the limited availability of fluent speakers able to participate in such experiments, as individual speakers even of the same language may display different reduction behaviour (Hanique et al., 2015). General reduction patterns in Lower Luga and adjacent areas had, however, been established prior to this experiment by Kuznetsova (2009, 2012a, 2016) on the basis of existing published sources as well as audio data on several dozens of speakers. It was observed that the degree of reduction increases from the north to the south of the Lower Luga area towards the Estonian language, which has been the most innovative in this respect and has completely lost reduced vowels. For example, the abovementioned varieties were graded from least to most susceptible to reduction in the following way: Kurkola Ingrian Finnish > Votic and Central Lower Luga Ingrian > South Lower Luga Ingrian > Siberian Ingrian/Finnish. Observed processes include qualitative and quantitative reduction, devoicing and speech elision, e.g., *püssü* ['pys:y] > ['pys:y̆] > ['pys:y̆̊] > [pysʲʷ:] > [pysʲ:] > [pys:] "rifle." While vowels still preserve their segmental status in the Lower Luga area, they turned into the consonantal features of labialisation and palatalisation in Siberian Ingrian/Finnish (Sidorkevič, 2013).

DOI: 10.1159/000494927

Table 1. Sociolinguistic data on the speakers

Variety code	Language	Dialect	Subdialect	Sex	Year of birth	Birth place	Recording place
IF	Finnish	South-Western	Kurkola Ingrian Finnish	F	1933	Hakaja	Sutela
V	Votic	Western	Luutsa	F	1928	Liivčülä	Liivčülä
CI	Ingrian	Lower Luga	Central	F	1927	Ropsu	Ropsu
SI1			South	M	1924	Vanakülä	Vanakülä
SI2				F	1932	Dal'n'aja Pol'ana	Dal'n'aja Pol'ana/ Narva
S	Ingrian/ Finnish (mixed)	Siberian variety: South Lower Luga Ingrian/ Rosona Ingrian Finnish		F	1950	Ryžkovo (Omsk region)	Tallinn

The role of the lexical and grammatical factor at the initial stages of reduction, predicted by the usage-based approaches (see 1.1), has also previously been noticed in the Finnic languages of Ingria and South-Western Finnish dialects. Grammatically conditioned vowel elision is claimed to have emerged earlier than purely phonetically conditioned elision (Laanest, 1980, pp. 73–74), and it occurred first of all in the most frequently used morphemes (Leskinen, 1973, p. 218). Specifically, lexically and grammatically conditioned vowel reduction has been attested in phonologically more archaic Ingrian Finnish and Soikkola Ingrian, while in more innovative Lower Luga Ingrian the conditioning is generalised to purely phonetic (Kuznetsova, 2016, pp. 9–11).

The present experimental study was designed following the patterns established in the abovementioned works and aimed at further clarifying the results obtained mainly from auditory impressions. All these languages share the same type and drift of reduction and differ just by its degree. Therefore, in this case it is possible to transpose this geographic variability along the north-south axis into the reduction progress along the time axis. All four languages are severely endangered: the number of speakers ranges from fewer than 10 to a couple of hundred (Kuznetsova et al., 2015; Sidorkevič, 2013). Therefore, the observed differences in production and categorisation of reduced vowels can hardly be attributed to the very fact of their endangerment.

The vowel inventories of these varieties contain low, mid and high vowels, front and back vowels, and labialised and non-labialised vowels. The systems in their most archaic variant in terms of non-initial vowel reduction, which can serve as a reference point for the processes described in the study, can be summarised as follows: *i:ī*, *ü:ǖ*, *u:ū*, *e:ē/ie*, *ö:ȫ/üö*, *õ:ȭ*, *o:ō/uo*, *ä:ǟ* and *a:ā*. The unrounded back vowel *õ:ȭ* is present only in Votic. The languages are characterised by significant prosodic differences between initial (stressed) and non-initial (unstressed) syllables. In certain varieties, the long initial mid vowels were raised to the diphthongs *ie*, *üö* and *uo*. Stems in all varieties are characterised by fronting vowel harmony within the domain of the root plus the following derivative and inflectional suffixes, as in Standard Finnish (*a*, *o* and *u* can occur in back-vowel stems, *ä*, *ö* and *ü* in front-vowel ones, and the "neutral" vowels *i* and *e* in both; for irregularities in Votic see, e.g., van der Hulst, 2018, pp. 176–178).

In general, short vowels can undergo reduction and loss at the end of a final or a non-final foot and in the second syllable of a trisyllabic foot (cf. examples from South Lower Luga Ingrian: *tütt***õ** [ˈtyt:ø̥̆] "girl," *avaht***ŭ** [ˈava·htŭ̥] "be_opened:PST:3SG," *sukk***ŭ***riš̄* [ˈsuk:ŭ̥ˌrˈis:] "sugar:IN," *lisäh***ṻ***mine* [ˈlisä·hˈẙ̆ˌmine·] "be_added:NMLZ," *ihm***ĭ***sed* [ˈihmĭ̥sed] "man:PL," *jāh***ṻ***tetā̈* [ˈjæ:hˈẙ̆teˌt:ä] "cool:IPS," *kisk***o***humine* [ˈkiskŏ̥huˌmine·] "be_torn_off:NMLZ"; for a detailed chart of reduction and loss positions see Kuznetsova, 2011, p. 189; 2012a, pp. 59–60). Reduction in non-initial long vowels is outside the scope of this study; for a general account see Kuznetsova (2016).

2.2 Methods of Data Collection and Analysis

In the phonetic experiment, open disyllables ending in the three corner types of vowel *a*, *i* and *u* (or *o*) after both voiced (*n*, *l*, *r*, *m* or *v*) and voiceless (*t*, *k*, *p*, *s* or *h*) singleton consonants were studied in the phrase-initial and the phrase-final position (3 vowels × 2 consonants × 13 words × 4 iterations × 2 positions = 624 tokens per sample). Most types of word-final combination of these vowels with consonants were covered. Based on existing phonological descriptions of these languages (Leppik, 1975; Kuznetsova, 2009; Markus & Rožanskij, 2011), one can argue that at least consonantal palatalisation might be stronger in front-vowel stems, and that geminates could be affected by it less than singletons. We, therefore, limited ourselves to singleton consonants and to stems with back and neutral vowels. Chosen stems were mostly morphophonologically back (the few front-vowel stems, which contained only the neutral vowels *i* and *e*, are underlined in Appendix 1).

The questionnaires were nearly identical (~5% variability) for all varieties, which share a substantial part of the lexicon. Words ending in *o* were taken instead of those with *u* in about one-third of cases. First, rounded vowels are much rarer in non-initial syllables than unrounded ones. Due to the endangered state of the varieties, it proved in some cases impossible to find words ending in the required combinations of *u* and a consonant that would be familiar to the speakers. Second, in the process of vowel reduction and loss in these varieties, the mid vowels *o*, *ö* and *e* are raised to *u*, *ü* and *i* (Mägiste, 1925, p. 80; Kuznetsova, 2012a, 2012b, 2016; see, e.g., *maito/maitu* "milk," *pudro/pudru* "porridge" and *viero/vieru* "wheel" in Appendix 1). Third, the loss of both *o* and *u* results in consonantal labialisation; thus, from this point of view, they are functionally similar.

The two phrasal positions were thought to be prosodically different enough to attest a wide range of phonetic variability in vowel realisations. Words in the phrase-initial position were pronounced in the context before the consonant *s*. The most typical position for complete vowel loss in these varieties is in sandhi before a following vowel. A position before a consonant was chosen because it allowed for subtler differences in the process of loss of different vowel qualities to be better identified. In the prevocalic context, where all vowels are nearly invariably lost in fast speech, these differences are neutralised. The data were recorded with a Zoom H4n digital recorder in the field, segmented and classified in Praat, and analysed in SPSS 11.5.0. The speakers had to translate the Finnic sentences with a carrier word from a phrase asked in Russian and repeat them 4 times. We subsequently counted the ratios of various types of vowel reflex within each pool of realisations along several scales. The most general binary scale included two main types: (1) "vowel" and (2) "loss." These were further divided into six subtypes in the following way:

"Vowel" (= "vocalic" realisations): (1) modal, (2) partially devoiced or (3) fully devoiced vowels

"Loss" (= "consonantal" realisations): (4) heavy segmental aspiration (>30–35 ms) after the consonant, (5) palatalised or labialised consonant or (6) complete vowel loss without any traces

For the first three speakers (IF, V and CI), a more detailed scale of variability in production was also used. These speakers showed low ratios of vowel loss, but still significant reduction in vowel quality. Their "vocalic" types were additionally assessed according to three independent scales:

Presence of strong laryngealisation: (1) yes or (2) no

Vowel quality: (1) full, (2) partially reduced or (3) completely reduced (to schwa)
Devoicing: (1) modal; (2) slightly aspirated vowels; vowels with (3) 10–30%, (4) 30–50%, (5) 50–70% or (6) 70–90% of devoicing; or (7) full devoicing

The devoicing scale is a more detailed variant of the six-type scale: type 1 of the latter includes devoicing types 1–3; type 2 includes devoicing types 4 and 5; and type 3 includes devoicing types 6 and 7.

Vowel reflexes were classified manually on the basis of the spectrographic data. Examples of main types of realisation are given in Figures 1 and 14–16. The "vocalic" types still preserved F1 and F2. If just one formant was present, the case was considered as "heavy segmental aspiration." Being shorter than 30–35 ms, such aspiration was seen as a consonantal feature of palatalisation or labialisation.

In a separate session from the phonetic experiment, we ran a parallel psycholinguistic test on how speakers categorise the reduced vowel reflexes. They were asked to write down in any preferred orthography the carrier words from the phonetic questionnaire the way they perceived them (~78 words). The task was formulated in Russian as: "Please write down a word for 'bird' in your language whatever way you prefer." If speakers noted that they did not know how to write in their language, the researcher emphasised her interest in the way how a person "feels" the word, not in the right orthography.

Three speakers used Cyrillic and three others (IF, SI1 and S) Latin letters; speakers SI1 and S, though, also sporadically used Cyrillic letters. For example, **lintu* "bird" could be typically written as *линту/lintu* or *линт/lint*. We did not give a multiple-choice task to the speakers, so as not to attract their attention to the final vowels. However, if speakers spontaneously noted that a word could be pronounced both with a vowel and without it, we counted these cases as two separate tokens. We counted the ratios of final vowel presence to loss for each speaker (sizes of the samples: IF = 78, V = 81, CI = 76, SI1 = 78, SI2 = 81 and S = 85).

Neither variety has a literary standard, so such a test provided a unique opportunity to observe more or less directly speakers' intuitions about the presence/absence of a vowel word-finally. At the same time, a classic perception test was not possible in those field conditions, given the advanced age and fragile health conditions of the subjects. The Russian language and the Finnic varieties belong to different families (Indo-European vs. Altaic), so the Russian tokens for carrier words were not expected to significantly influence the outcome of the test. Moreover, both the Cyrillic and the Latin mediating orthographies rely on the phonemic principle of encoding, and so they automatically prompted subjects to reflect in writing whether there was any vowel word-finally or not.

In some cases, a more detailed scale was used for this categorisation test:
"Vowel": (1) full vowels or (2) reduced vowels
"Loss": (3) retention of consonantal palatalisation or labialisation, or (4) zero

Palatalisation was coded by speakers with the use of the Russian "soft sign" "ь." The results of the categorisation of palatalisation and labialisation should be considered tentative, as the Russian orthography does not have a corresponding sign for labialisation. The latter was depicted only by the Siberian speaker as *(o)* or *(u)* in parentheses after the consonant, while she explicitly claimed the absence of final vowels. The observed asymmetry in the depiction of the two features might be partially influenced by this orthographic disparity. Reduced vowels were rendered by some by means of the Russian "*ы*" (high unrounded mid vowel).

Fig. 1. Examples of reduction in vocalic quality. **a** Full modal V (**hāpa* [ha:pa] "aspen") MO. **b** Laryngealised V (**velka* [veɫka̰] "debt") MO. **c** Schwa (**āpa* [a:pə] "aspen") PK. **d** Slightly aspirated V (**kumpa* [kumpa^h] "which of the two") MO. **e** Partially devoiced V (**āpa* [a:pa̤] "aspen") PK. **f** Devoiced V (**āra* [a:rḁ] "branch") PK.

(For figure see next pages.)

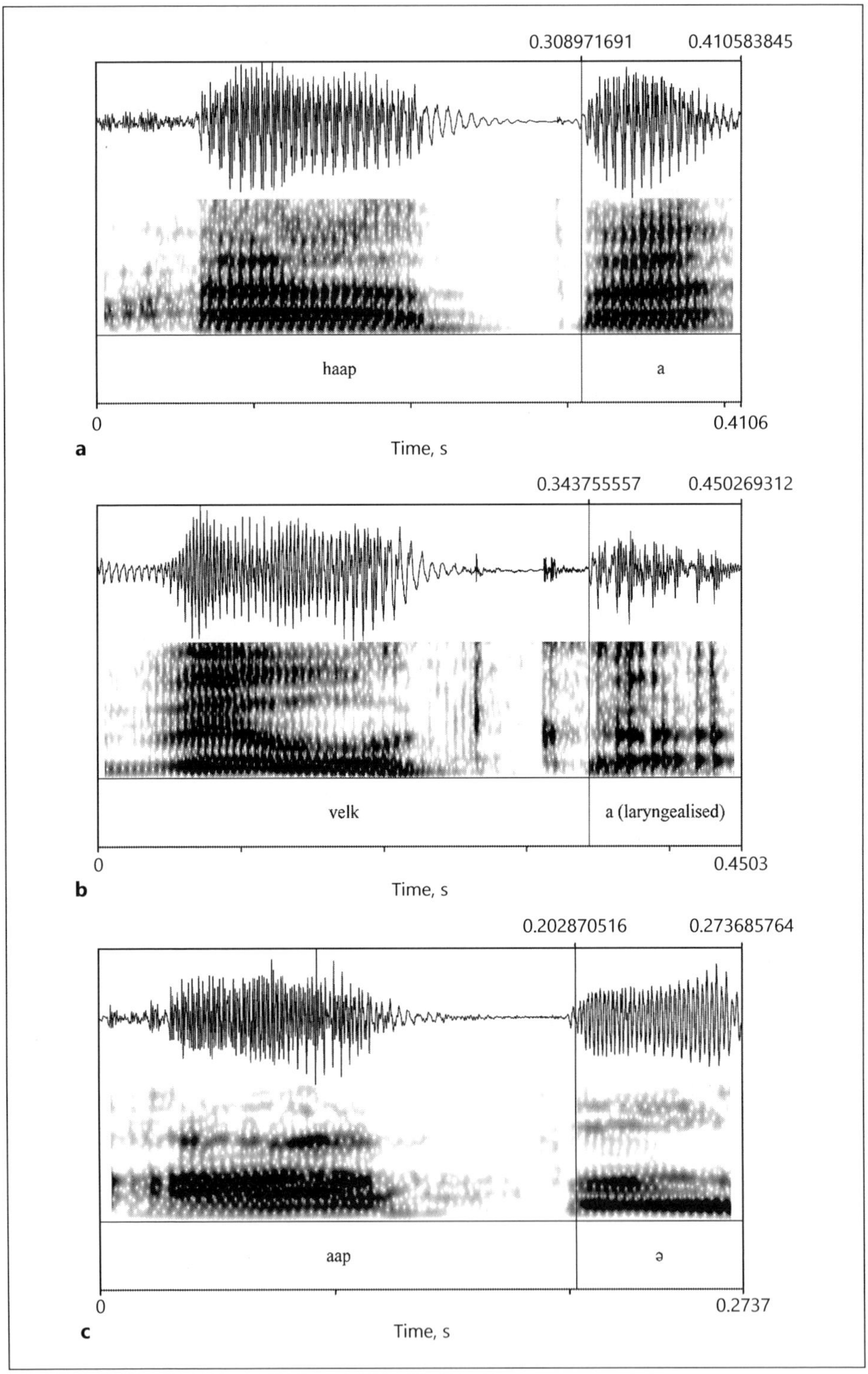
0.308971691
0.410583845
haap
a
0
0.4106
a
Time, s
0.343755557
0.450269312
velk
a (laryngealised)
0
0.4503
b
Time, s
0.202870516
0.273685764
aap
ə
0
0.2737
c
Time, s

1

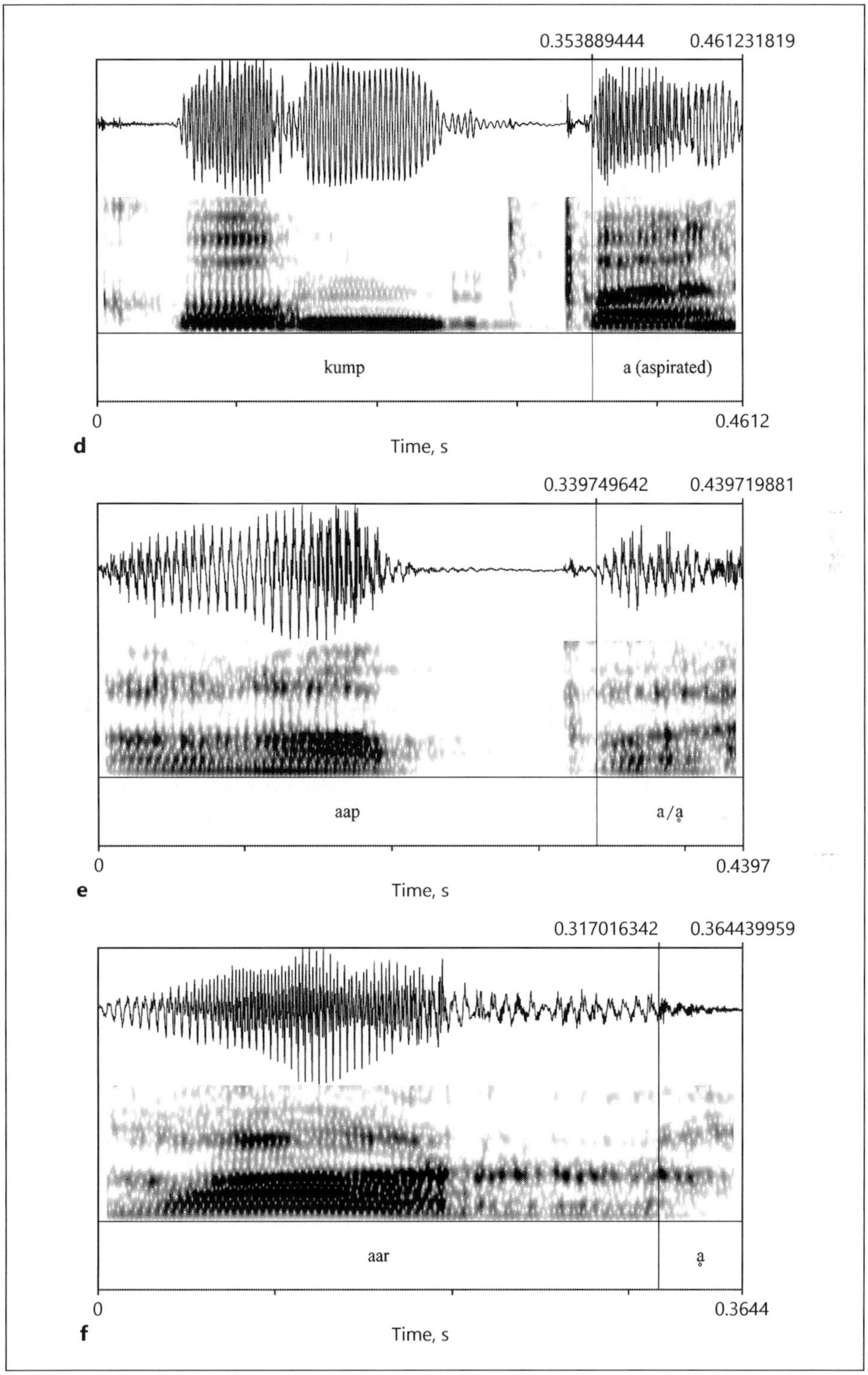
0.353889444
0.461231819
kump
a (aspirated)
0
0.4612
d
Time, s
0.339749642
0.439719881
aap
a/ḁ
0
0.4397
e
Time, s
0.317016342
0.364439959
aar
ḁ
0
0.3644
f
Time, s

1

3. Results and Discussion

The results show a robust correlation between production and categorisation. The general structure of the category prototypes (Rosch, 1978) was the same in phonetic realisation and phonemic representation at each of the three stages of vowel loss observed (see Fig. 11). At stage 1 (IF, V and CI), "vocalic" realisations composed more than 90% of the sample, which correlated with their only one robust mental prototype [+SEGMENT]. At stage 2 (SI1 and SI2), there was a roughly 50/50 split both between "vocalic" and "consonantal" realisations, on the one hand, and [+SEGMENT] and [–SEGMENT] categorisations, on the other hand. At stage 3 (S), with vowel loss in >70% of the cases, only one category prototype [–SEGMENT] prevailed. These results clarify the findings by Kuznetsova (2016), where less phonetic reduction was expected for Kurkola Ingrian Finnish and more for Central Lower Luga Ingrian, respectively.

Below, we address individual features of production and categorisation and summarise the tendencies at each stage. In the General Discussion (section 4), we outline main trends in the loss of vowel quality and main differences between the six speakers and the three vowel types. Phonetic differences across positions, consonantal types and individual words and nuances concerning vowel duration largely remain outside the scope of this paper. The differences across phrasal positions and after voiced versus voiceless consonants were indeed noticeable in terms of the percentage of vowel loss, duration, and quality. Vowels expectedly manifested much more devoicing after voiceless consonants. Initial phrasal position was primarily characterised by strong qualitative reduction (apparently triggered by extremely reduced duration), while final position exhibited more devoicing. Vowel duration divided speakers into two groups: (1) stage 1 – short vowels (90–100 ms) phrase-finally and reduced vowels (<80 ms) phrase-initially, and (2) stages 2 and 3 – reduced vowels (<80 ms) in both phrasal positions.

3.1 Stage 1: Ingrian Finnish, Votic and Central Lower Luga Ingrian

The samples at stage 1 of reduction belong to three different languages: Finnish, Votic and Ingrian. Even if similar in the general structures of distribution, they exhibit slightly different configurations of vowel loss in realisation and categorisation (see the percentages of loss in Fig. 5–7). Ingrian Finnish represents the most conservative variety, and Central Lower Luga Ingrian the most innovative one, with Votic in the middle. In all three samples, the vowel **u*[1] reveals exactly the same pattern, being the most conservative of all the vowels: full preservation in mental categorisation and just 1–2% loss in production. What differs across the varieties in question is the configuration of the vowels **a* and **i*. In Ingrian Finnish, **i* is the most innovative in terms of both production and categorisation, while in Central Lower Luga Ingrian, it is the vowel **a*. In Votic, the production pattern corresponds to the one found in IF (**i* is the most innovative and **a* is as conservative as **u*), while the categorisation rather resembles that of CI, where **a* is more innovative than the other vowels.

[1] Hereafter, the original etymological vowel qualities, whose reflexes are studied in the experiments, are marked with an asterisk.

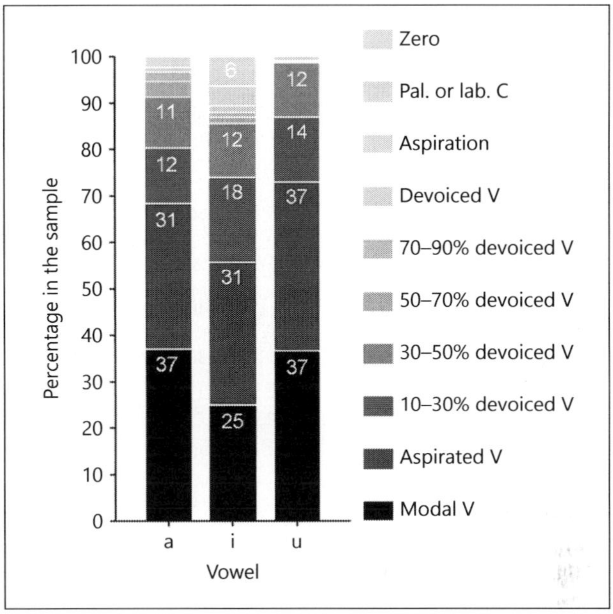

Fig. 2. IF (Ingrian Finnish speaker): vowel devoicing. pal., palatalised; lab., labialised.

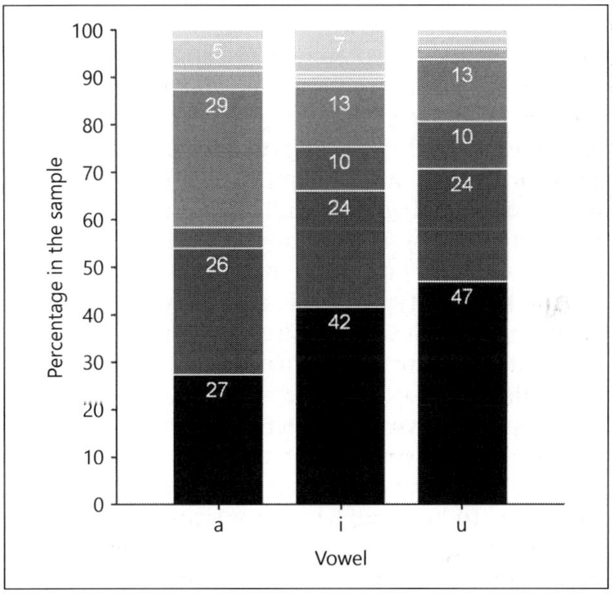

Fig. 3. V (Votic speaker): vowel devoicing. For the legend explaining the types of devoicing, see Fig. 2.

A more detailed look at vowel devoicing (Fig. 2–4) and reduction of quality (Fig. 5–7), as well as lexical considerations (Table 4), clarify the possible reasons for these differences. Vocalic segments are still largely present at stage 1, but their quality is reduced along three dimensions: aspiration, centralisation and laryngealisation (cf. Klatt & Klatt, 1990; Laver, 1994, pp. 189–191; Barnes, 2006,

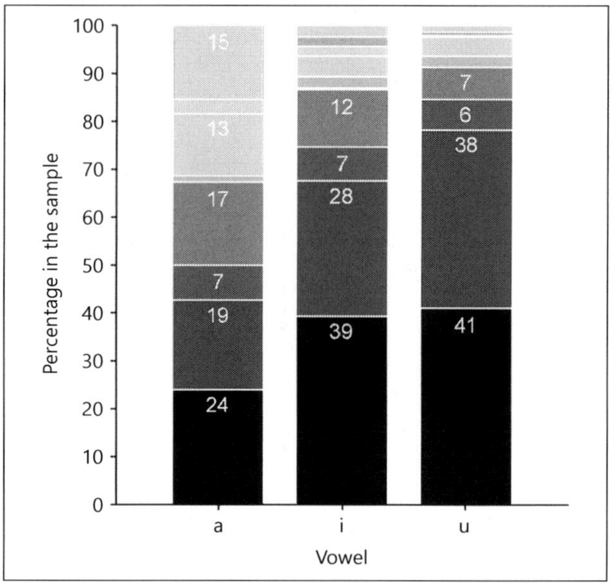

Color version available online

Fig. 4. CI (Central Lower Luga Ingrian speaker): vowel devoicing. For the legend explaining the types of devoicing, see Fig. 2.

pp. 114–150). Completely non-aspirated variants, in fact, accounted for just about half of those realisations which were considered modal according to the six-type scale (see Fig. 2–4). In total, partially or fully aspirated and devoiced allophones overwhelmingly prevailed over the "clear" modal ones even at stage 1. The percentage of non-aspirated modal allophones is in negative correlation with the percentage of complete loss in production in nearly all the cases (apart from **a* in Votic). The prototypical realisations (type 6: "zero") of the new category [–SEGMENT] are therefore gaining strength in production first of all at the cost of the prototypical realisations (type 1: "modal non-aspirated vowel") of the old category [+SEGMENT]. The belt of intermediate types preserves roughly the same structure for all the three vowels within each speaker and just slides down the scale.

Qualitative vowel reduction reveals quite a different picture (Fig. 5–7). Noticeable differences in the structure of phonetic variability appear between vowel types but not across speakers. Phonetic reasons for these differences are, therefore, to be sought in the articulatory and perceptual properties of vowel qualities rather than in other factors. In all three samples, *a* has undergone an extremely strong reduction to schwa (around 65% of complete schwa realisations and <30% of full vowels), and at the later stages of reduction, it is realised as schwa almost invariably. In contrast, **i* was the vowel most likely to retain its quality (around 80% of non-reduced allophones). The vowel **u* occupied an intermediate position, with about half of its realisations being non-reduced. In quality assessment, a more detailed scale for categorisation was used (see 2.2). The category of a reduced vowel rendered via Russian "*ы*" occurred only in the vowel **a* of the Votic speaker.

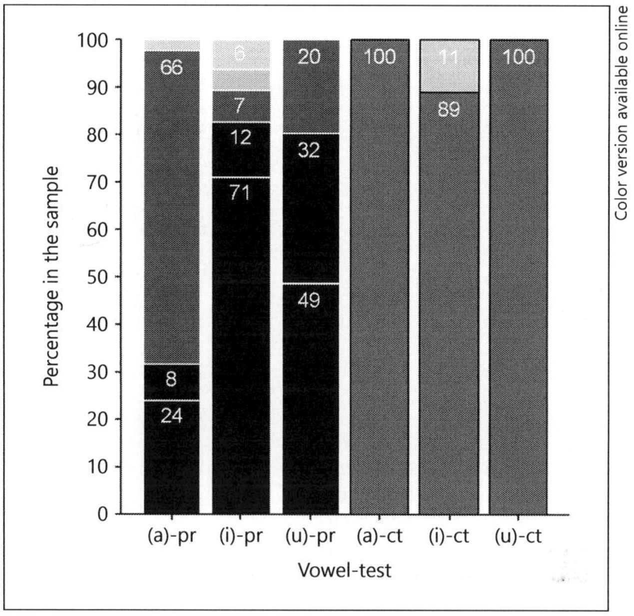

Fig. 5. IF (Ingrian Finnish speaker): qualitative reduction. pr, production; ct, categorisation. For the legend explaining the types of qualitative reduction, see Fig. 6.

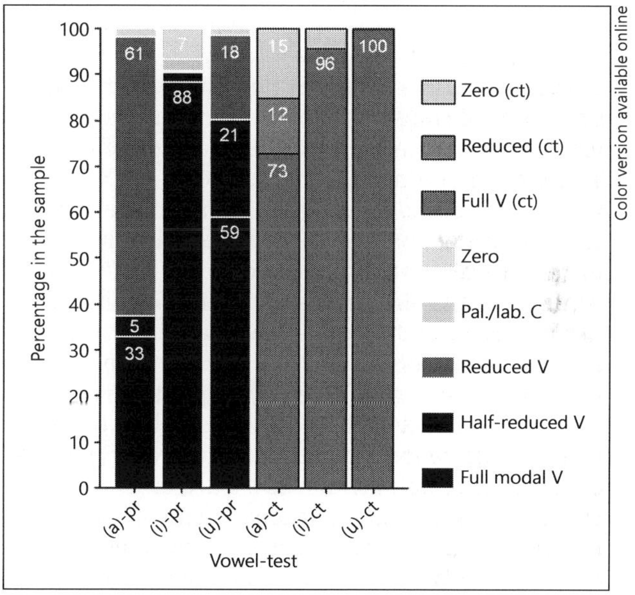

Fig. 6. V (Votic speaker): qualitative reduction. pal., palatalised; lab., labialised; pr, production; ct, categorisation.

Qualitative reduction and devoicing manifested very differently, sometimes with opposing distributional patterns. We explored which of the two correlated better with the percentages of vowel loss in production and categorisation.

In Ingrian Finnish, the devoicing structure is the same for all vowel types, so it cannot be a factor conditioning the differences in their loss. Qualitative re-

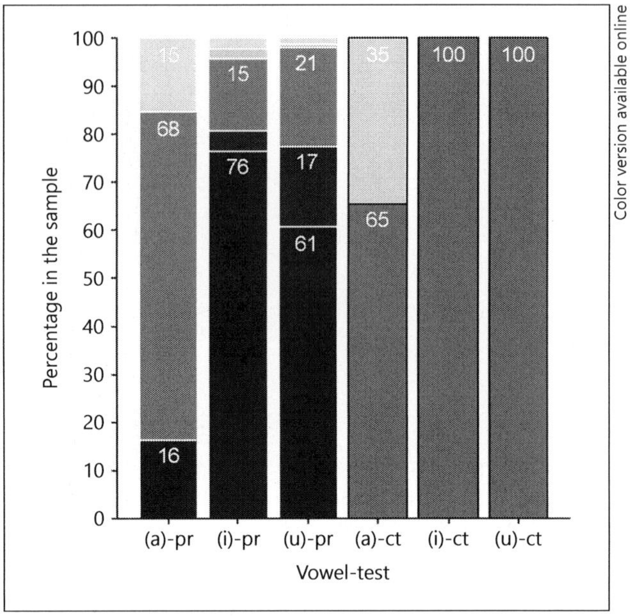

Fig. 7. CI (Central Lower Luga Ingrian speaker): qualitative reduction. pr, production; ct, categorisation. For the legend explaining the types of qualitative reduction, see Fig. 6.

duction, in turn, correlates negatively with loss in production and categorisation. The only obvious phonetic factor correlating with the level of loss is, therefore, the type of vowel itself, as the level of loss in **i* is higher than in the other vowel types (for non-phonetic factors see 3.4).

In Central Lower Luga Ingrian, on the other hand, loss in production and categorisation positively correlates with the level of devoicing in all cases, and in **a*, also with qualitative reduction. One could say that in **a*, devoicing and centralisation reinforce each other as phonetic drivers for reduction, resulting in a relatively high percentage of loss in production (19%) and even more so in categorisation (35%). We will see later that in Lower Luga Ingrian, it is indeed vowel devoicing, reinforced by qualitative reduction, that is the primary driving force of loss, especially of the drastic loss of **a* from speakers' awareness and production. Devoicing with quality preservation leads to the rise of phonemic consonantal palatalisation as a trace of **i*.

The Votic speaker presented a mixed strategy between these two. As in IF, loss in the production and categorisation of **i* did not correlate with devoicing and negatively correlates with centralisation. At the same time, the configuration for **a* resembled that found in CI, although the equation was not perfect. The level of loss in categorisation positively correlated with those of qualitative reduction and devoicing. At the same time, the level of loss in production correlated with all three negatively. In other words, even if the speaker centralised and devoiced **a*, this did not lead to an increased drop of this often voiceless schwa from her production. In fact, Votic **a* was the only vowel in the stage 1 speakers which showed a clear negative correlation between the levels of loss in production and categorisation. The loss of reduced **a* had already

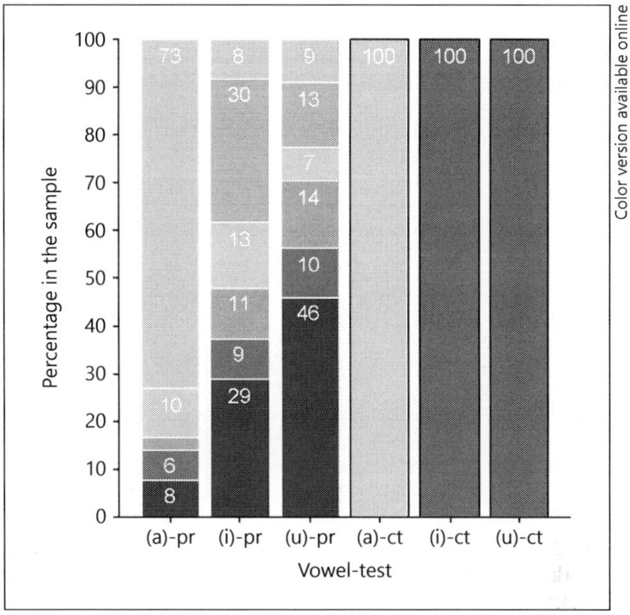

Fig. 8. SI1 (South Lower Luga Ingrian speaker 1): vowel devoicing. pr, production; ct, categorisation. For the legend explaining the types of vowel devoicing, see Fig. 9.

started in the mental categorisation but not yet in the production of the Votic speaker, so she manifested also a more conservative production pattern for **a* than the CI speaker. She was the only one who was aware of the qualitative reduction of **a* among the stage 1 speakers. This awareness might be related to the presence of the unrounded mid back vowel *õ* in Votic, unique among these varieties (see 2.1). The Votic speaker might have identified the schwa with this *õ*.

In IF and V, final vowels also underwent laryngealisation (27% of the IF tokens and 15% of the V tokens). In the CI speaker, this process was not attested. Summary laryngealisation patterns (IF + V) across the vowel types broadly correlated with the patterns for qualitative reduction. The vowel **i* tended to be the most conservative (17% of laryngealisation), and **a* the most innovative (26%), with **u* in the middle (19%), although these were not strong tendencies.

3.2 Stage 2 (South Lower Luga Ingrian) and Stage 3 (Siberian Ingrian/ Finnish)

The speakers at stages 2 and 3 of reduction (Fig. 8–10) manifested a continuation of the same tendencies, especially CI. The speakers at stage 2 belonged to the same variety (South Lower Luga Ingrian), but the male speaker SI1 was not a typical one. He used to be a community manager and a local cultural leader and has a notably higher level of linguistic awareness than others. In his notebooks, one can find texts and words in his own variety in an orthography created by himself, reasonings for choices of orthography, texts in other Finnic languages copied from published sources, and etymological comparisons between cognate Finnic words.

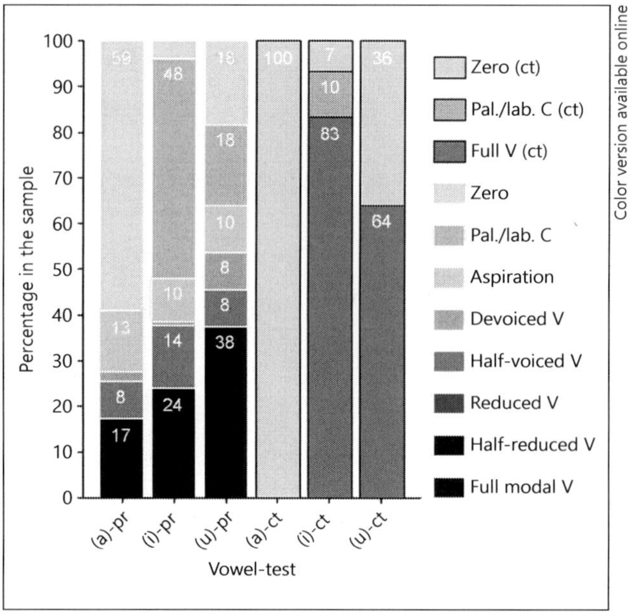

Fig. 9. SI2 (South Lower Luga Ingrian speaker 2): vowel devoicing. pal., palatalised; lab., labialised; pr, production; ct, categorisation.

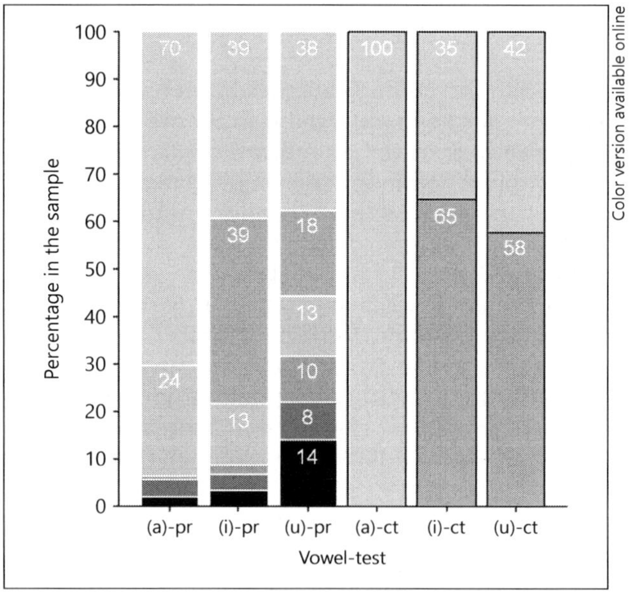

Fig. 10. S (Siberian Ingrian/Finnish speaker): vowel devoicing. pr, production; ct, categorisation. For the legend explaining the types of vowel devoicing, see Fig. 9.

The main reasons for the differences observed between SI1 and SI2 could be attributed to these specific characteristics of SI1. He has a more innovative production pattern for *a, and more conservative ones for *i and *u, than SI2. Categorisation suggests a clue to the origins of this difference. In SI1, it was the most consistent of all six speakers (apart from the Siberian speaker, where the

DOI: 10.1159/000494927

sound change process had already reached the terminal stage). He categorised the final *a > ə always as zero, although he was actually not as consistent regarding schwas in non-final positions, where he often used "ы" (Kuznetsova, 2012b), much as the Votic speaker did. Seemingly, this is the closest perceptual Russian correlate of schwa for speakers of these local languages. Final *i and *u, on the other hand, were always perceived as vowels, though the Siberian speaker was aware of their reduced character and called them "half-vowels." He seemed to target these mental categorisations in his pronunciation consciously, and his percentage of loss was correspondingly higher for *a and lower for *i and *u when compared to the otherwise linguistically very close speaker SI2. He was obviously not able to gain full control over his production, though, and his pattern of loss for the three vowels still has a scalar shape similar to that of other Ingrian speakers (CI and SI2).

In SI2, the phonemic categorisation (as zero) was consistent only for the vowel *a, which had reached the critical threshold for complete loss. Interestingly, *i and *u showed reverse patterns of loss in her production versus categorisation, which is apparently rooted in the acoustic and perception properties of these two vowels. In general, at stages 2 and 3, one observes a robust cluster of palatalisation for *i in all speakers (~40% of phonetic production). The vowel *u also manifested a visible cluster of labialisation, completely absent at stage 1, but it accounted for only 15% of its phonetic production. If one adds the clusters of strong aspiration and palatalisation/labialisation to the "vocalic" realisations of vowels, the distributional patterns in production of *i and *u of both SI speakers match those of categorisation much more closely. It seems that at the intermediary stage 2, the "consonantal" reflexes of the vowels *i and *u, which give colour to consonants, still correspond rather to [+SEGMENT] in mental categorisation. Especially in the case of *i, one could argue that both SI1 and SI2 still perceived and targeted, more or less consciously, the full vowel. This might be linked to a robust salience of *i-reflexes in both perception and articulation (see 4.2). The speakers succeeded in reaching a vowel in only about half of the cases, though, ending up with a more or less aspirated palatalised final consonant in the other half instead. The lower perceptual salience of *u can be seen in its relatively innovative categorisation by SI2, which was, in general, not as systematic as that by SI1. For the even less perceptually salient *a, however, a relatively robust cluster of consonantal aspiration did not prevent complete loss from categorisation, as *a does not colour consonants.

In the Siberian speaker at stage 3, we saw the next step of the same processes. Here, all the vowels had already reached the critical threshold for loss in production in order to be lost from mental categorisation. Judging by all three samples from stages 2 and 3, one could estimate this threshold at about 70%. Categorisation becomes innovative for vowels which have reached it, while their production still lags behind. In the Siberian speaker, the structure of phonetic loss for *i already closely followed that of *a; that is, the middle step of the "ladder-like" pattern flattened. We still saw a more conservative production of *u with respect to the other vowels, though. The phonetically conservative nature of *u, observed in all the speakers at stages 2 and 3, cannot be explained by its categorisation properties and should apparently be attributed to general physiological factors (acoustics, articulation, perception and storage in the memory;

see 1.2 and 4.2). At the same time, while in the Siberian speaker the palatalisation cluster was still as big as the cluster of complete loss, the labialisation cluster was already twice as small as the latter. One might hypothesise that while palatalisation could still have a chance of being preserved as a phonemic feature in these languages, labialisation has already lost the historical sound change battle, even though the Siberian speaker still perceived its presence. The aspiration cluster as a reflex of **a* was especially robust in the Siberian speaker, but as at stage 2, this did not affect its perceived complete loss.

3.3 Statistical Tests on the Main Findings

The main results of the study are corroborated by statistics. We compared the means of "vowel" (= 0) versus "loss" (= 1) in the overall production and categorisation, across the vowels, the speakers, the vowels in speakers and the stages of reduction. One-way ANOVA and Levene's F showed a highly significant difference ($p < 0.001$) in all cases apart from the first comparison. The overall production ($n = 3{,}744$, $M = 0.36$, SD = 0.481, SE = 0.008) and categorisation ($n = 479$, $M = 0.35$, SD = 0.479, SE = 0.022) did not differ ($F(1, 4{,}221) = 0.12$, $p = 0.729$, Levene's $F = 0.501$, $p = 0.479$), which supports the general correlation in production and perception of vowel loss (the difference in SE can be explained by the unequal size of the groups). For the other cases, we ran two post hoc tests for pairwise within-group comparisons in big samples of unequal size and variance: Tamhane's *T2* (*T*; more conservative) and Games-Howell (GH; more liberal) at the 95% confidence interval. The few differences between them are reported below as *T*/GH.

There was a significant effect of stage on the level of loss ($F(5, 4{,}217) = 654.27$, $p < 0.001$). Both post hoc tests showed no difference between production and categorisation at stage 1 (MD = –0.02, SE = 0.018, $p = 1$), a difference at stage 2 (MD = 0.15, SE = 0.042, $p = 0.004$) and a highly significant difference at stage 3 (MD = –0.16, SE = 0.015, $p < 0.001$). In other words, both production and categorisation are still conservative at the first stage, then production becomes significantly more innovative, which leads to the shift in categorial analysis: categorisation becomes significantly more innovative and drives the loss at the terminal stage.

The overall results for the three vowels showed a highly significant difference ($F(2, 4{,}220) = 77.49$, $p < 0.001$), as each vowel has its unique configuration of loss in production and categorisation (Table 2; Fig. 12). The post hoc tests on these two aspects analysed separately showed that both the production and the categorisation of **u*, as well as the categorisation of **i*, did not differ and were conservative. At the same time, the more innovative production of **i* manifested a relatively significant difference from these three, being closer to the even more innovative production of **a*. The latter was insignificantly more conservative than the production of **a*. In sum, **u* was conservative and **a* innovative in both aspects, while **i* was conservative in categorisation and intermediate in production.

The overall results for the six speakers lacked any difference for IF and V (MD <0.01, $p = 1$), but CI differed from these two (CI-IF: MD = 0.05, $p = 0.003$ *T*/0.002 GH; CI-V: MD = 0.05, $p = 0.002$). CI manifested a slightly more advanced substage of reduction inside stage 1, where **a* overtook **i* as the leader of loss. At

Table 2. Mean differences between production and categorisation in vowels

Vowel (test)	i (pr)	u (pr)	a (ct)	i (ct)	u (ct)
a (pr)	0.07** ($p = 0.006$)	0.21***	−0.11 ($p = 0.086$ T/ 0.065 GH)	0.21***	0.22***
i (pr)		0.14***	−0.18***	0.14** ($p = 0.002$)	0.15** ($p = 0.002$ T/ 0.001 GH)
u (pr)			−0.32***	0.0	0.01
a (ct)				0.32***	0.33***
i (ct)					0.01

pr, production; ct, categorisation. ** $p < 0.01$, *** $p < 0.001$.

stage 2, SI1 did not significantly differ from SI2 (MD = −0.06, $p = 0.262$). All other differences between speakers were highly significant ($p < 0.001$). Production and categorisation are further analysed across speakers in Table 3 (see also Fig. 11, 13). At stage 1, both production and categorisation by IF and V, as well as categorisation by CI, did not differ and were conservative. Production by CI was slightly more innovative: it showed moderate-to-weak differences from production by IF and V (but not from categorisation by CI). Production and categorisation by S highly differed from everything else and from each other (her production was significantly more conservative than her production due to the categorial shift at stage 3). At the intermediate stage 2, production and categorisation by SI1 and production by SI2 did not show significant differences. However, categorisation by SI1 stands out against all effects in Table 3. The peculiarity of categorisation by SI1 is likely linked to the speaker's unusual linguistic awareness and full systematicity in transcription (**a* as zero, **i* and **u* as vowels; see 3.2).

3.4 The Lexical Factor in Reduction at Stage 1

A lexical factor effect at the initial stage of reduction (see 1.1, 2.1) was observed also in the present data, with nuances concerning vowel type, speaker and the correlation between production and categorisation. Table 4 shows the data on the lexical distribution of the cases of loss at stage 1. In column 1, the total number of lexical words in each sample is given. Column 2 provides the number of lexemes in which at least one token of "loss" in production was attested, their percentage in the sample for each speaker, and their distribution across the **a*, **i* and **u/o* types. The vowel types are arranged in the parentheses from those with the highest number of words exhibiting loss to those with the lowest number exhibiting loss. Column 3 cites only those words in which more than half of the tokens showed loss (i.e., $n > 4$; the exact number of such tokens is given in parentheses for each word). Column 4 summarises the number of words from column 3 and their percentage in each sample. Finally, in column 5, the ratio between the numbers of words in columns 4 and 2 is calculated, providing an idea of the lexical compactness of the distribution of vowel loss.

Table 3. Mean differences between production and categorisation by each speaker

Speaker	V (pr)	CI (pr)	SI1 (pr)	SI2 (pr)	S (pr)	IF (ct)	V (ct)	CI (ct)	SI1 (ct)	SI2 (ct)	S (ct)
IF (pr)	0.0	−0.05(*) (p = 0.058 T/0.042 GH)	0.5***	0.55***	0.8***	0.01	0.03	0.07	0.3***	0.45***	0.95***
V (pr)		−0.05* (p = 0.013 T/0.011 GH)	0.51***	0.56***	0.8***	0.0	0.03	0.08	0.3***	0.45***	0.96***
CI (pr)			0.46***	0.4***	0.75***	0.06	0.02	0.02	−0.25** (p = 0.001)	0.4***	0.91***
SI1 (pr)				0.05	0.29***	0.51***	0.48***	0.43***	*0.2 (p = 0.042 T/0.03 GH)	0.06	0.45***
SI2 (pr)					0.24***	0.56***	0.53***	0.48***	**0.25 (p = 0.002 T/0.001 GH)	0.11	−0.4***
S (pr)						0.81***	0.77***	0.73***	0.5***	0.35***	−0.16***
IF (ct)							−0.04	−0.08	−0.31***	−0.46***	−0.96***
V (ct)								−0.04	**(*)0.27 (p = 0.001 T/<0.001 GH)	−0.42***	−0.93***
CI (ct)									*0.23 (p = 0.047 T/0.034 GH)	−0.38***	−0.88***
SI1 (ct)										−0.15	−0.65***
SI2 (ct)											−0.51***

IF, Ingrian Finnish; V, Votic; CI, Central Lower Luga Ingrian; SI, South Lower Luga Ingrian; S, Siberian Ingrian/Finnish; pr, production; ct, categorisation. * $p < 0.05$, ** $p < 0.01$, *** $p < 0.001$.

In IF and V, the level of the lexical compactness of loss was 4 times higher than in CI. The vowel loss in these speakers was concentrated in very few frequent basic words, while in CI the lexical dispersion of loss was much higher. It is remarkable that in IF and V, the loss in frequent words concerned only the vowel **i*. In Soikkola Ingrian, the other still existing Ingrian dialect which is about as archaic as IF from the point of view of reduction (Kuznetsova, 2016), the same type of **i*-loss in frequent words became lexicalised. For example, the following words in our questionnaire do not have the final **i* in Soikkola Ingrian: *pēn/pīn* "small," *ūs* "new," *sūr* "big," *laps* "child" and *nōr/nūr* "young." Grammatical morphemes (even more frequent elements of the language) of Ingrian Finnish, Votic and Soikkola Ingrian manifest such grammaticalised loss for both **i* and **a*. In CI, a speaker using a more innovative variety where reduction is already conditioned phonetically, **a* became the reduction leader.

These differences in production find a parallel in categorisation (see Table 4, column 6; words encoded with V/Ø were cited by the respective speakers as having both a variant with a vowel and one without it). In IF the loss was perceived only in **i*-words, in V it was perceived both in **i*- and in **a*-words, and in CI it was perceived only in **a*-words. The number of lexical items with perceived

Table 4. Lexical specification of vowel loss at stage 1

Speaker (*N*)	Lexemes exhibiting phonetic loss, *n*	Lexemes with phonetic loss *n* > *4* (>50%)	Lexemes with phonetic loss >50%, *n*	Lexical compactness of loss	Lexemes which exhibited loss in categorisation (by glosses)
IF (78)	**12** (15.3%) (*i, a*)	*pien***i** "small" (6), *ūs***i** "new" (8)	**2** (2.6%)	**0.17**	**i**: small, child (*i/Ø*), elk
V (81)	**10** (12.3%) (*i, a, u/o*)	*sūr***i** "big" (6), *pēn***i** "small" (6)	**2** (2.5%)	**0.2**	**i**: big (*i/Ø*); **a**: dog, change:IMP, barley, nail, bath broom
CI (76)	**21** (27.6%) (*a, i, u/o*)	*nāgl***a** "nail" (6)	**1** (1.3%)	**0.05**	**a**: nail, leg, which (of the two), black (*a/Ø*), change:IMP, debit, twig, bath broom, broom, class, skinny

IF, Ingrian Finnish; V, Votic; CI, Central Lower Luga Ingrian.

loss also increased from IF to CI. It is worth noting that the match between production and categorisation was close in the statistical sense, but not in the lexical one. The examples in Table 3 show that in each speaker's production and perception, the trends for reduction and loss correlate better across the vowel types than across the concrete lexical items. This might provide support for distributional learning of phonological categories, which happens relatively independently of individual lexemes and minimal pairs (see 1.1).

4. General Discussion

4.1 General Course of Vowel Reduction and Loss

Our study was restricted to two phrasal contexts (in the production part) and three types of vowel, and only one speaker was taken for each variety apart from one. Even if limited by these and other methodological restrictions, the results revealed a stable correlation of frequency distributions in production and perception across all six speakers. The latter represented three main stages of vowel reduction and loss in the Finnic languages of Ingria. This correlation of the internal structure of categorical representation to the structure of production is probably best explained by the adaptive hybrid models of mental storage, which suggest the internal clusterisation of exemplars within a category (Gureckis & Goldstone, 2008; Kirby, 2010, pp. 34–37).

The main vectors summarising the general course of vowel reduction and loss in the Finnic languages of Ingria and the differences across speakers and vowel types are represented in Figures 11–13 (the mean values for each test on a scale between 0 = "vowel" and 1 = "loss" and the SE bars are given; red stands for production and grey for categorisation). The dashed horizontal lines are drawn at 70% loss and at 70% preservation of the segments, which appeared to

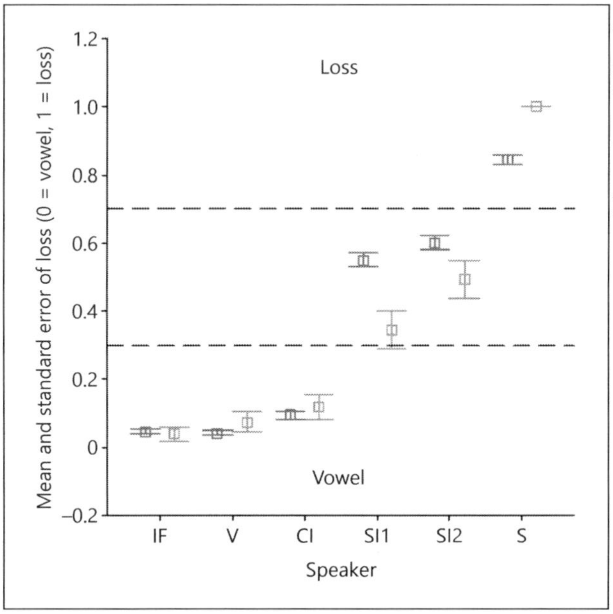

Color version available online

Fig. 11. Summary: speakers. Red stands for production and grey for categorisation. IF, Ingrian Finnish; V, Votic; CI, Central Lower Luga Ingrian; SI, South Lower Luga Ingrian; S, Siberian Ingrian/Finnish.

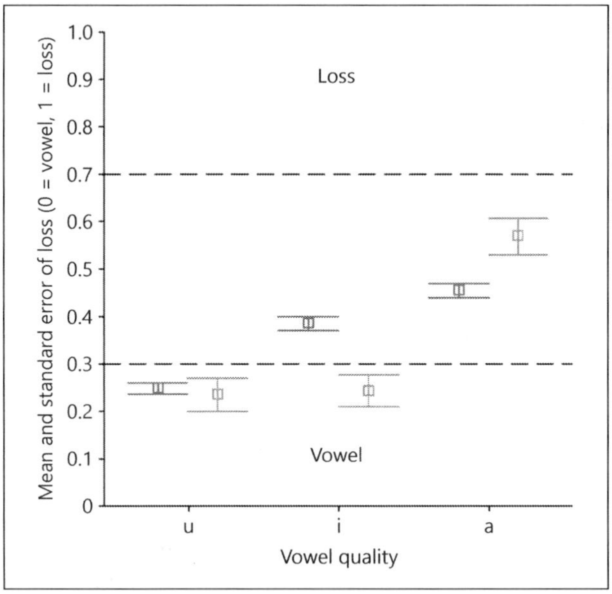

Color version available online

Fig. 12. Summary: vowels. Red stands for production and grey for categorisation.

be important thresholds for the stages of reduction and loss and changes in categorisation.

Indeed, one can observe the three stages of reduction described in the paper divided by these thresholds in Figure 11. At stage 1, production and categorisation are closely matched. As discussed in 3.4, vowel reduction and especially loss

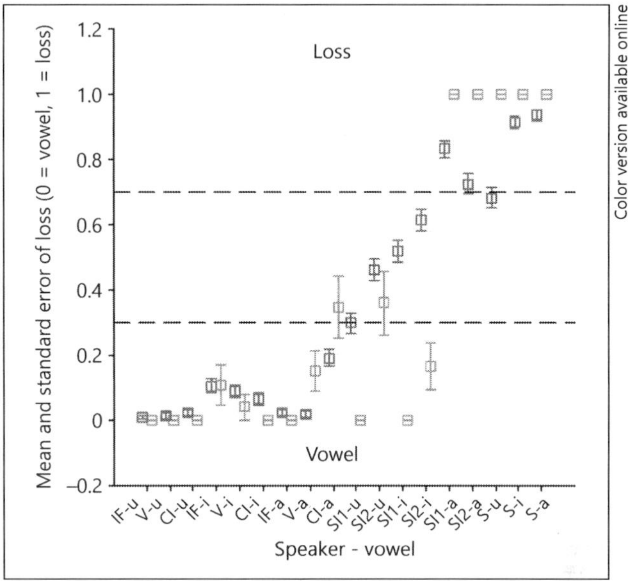

Fig. 13. Summary: speakers and vowels. Red stands for production and grey for categorisation. IF, Ingrian Finnish; V, Votic; CI, Central Lower Luga Ingrian; SI, South Lower Luga Ingrian; S, Siberian Ingrian/Finnish.

at this initial stage (speakers IF and V) is linked to a large extent to certain frequent elements (frequent lexemes, grammatical markers). At the same time, the correlation between production and perception in each vowel of each speaker is in general closer in the statistical sense than across concrete lexemes. The learned phonological category looks more like a sum of the distributional properties of phonetic variants in production, abstracted from particular lexical words, in line with the distributive learning hypotheses (see 1.1).

By the middle phase of loss, the novel stimuli have been accumulated and spread through the phonetic system, so reduction and loss are conditioned phonetically rather than lexically or grammatically. Categorisation remains more conservative at the first two stages, that is, more of the new category is produced than perceived. Speakers at stage 2 still often seem to target the old category in pronunciation, especially in the vowels which give colour to consonants, but achieve it only partially. The mechanism of reduction with a more conservative categorisation than realisation is linked to the automatisation of execution in production, is phonetically gradual and can likely take a considerable amount of time. If the old category ([+SEGMENT] in our case) still keeps 70% or more of realisation, the formation of the new category [–SEGMENT] is not yet perceived by speakers (stage 1). If both categories are pronounced in about 50% of cases, the categorisation is also split about 50/50 between the perceived presence and absence of a vowel (stage 2). When the new category has arrived at more than 70% of realisations (stage 3 and some vowels at stage 2), the crucial jump in categorisation has happened. The pattern is drastically reversed: the old value is no longer perceived, while it is still partially maintained in production. Reduction and loss at the stage of a more innovative categorisation imply a drop of low-salience meaningless parts, which is sometimes distinguished from the automatisation-based mechanism (see 1.2). At the same time, it might still mean automatisation in production, now of the new category rather than of the old one.

Major differences in production and categorisation observed between the three corner vowels are summarised in Figure 12. Each of the three vowels manifested its own combination of production and categorisation values. The vowel **u* turned out to be conservative both in production and categorisation, and in total the most conservative among these vowels. The vowel **i* had an intermediate position, with an overall categorisation as conservative, as in **u*, but production nearly as innovative as **a*. This innovativeness in loss is actually accompanied by the formation of a robust cluster of consonantal palatalisation (see 4.2). The vowel **a* was the most advanced in terms of loss, and here, in turn, categorisation was more innovative than realisation. This is obviously linked to the its extremely strong qualitative reduction and due to the fact that it does not leave any colour to the consonants. In general, the more the vowels were reduced to schwa, the less their presence was perceived.

Our results actually showed that the two main patterns of vowel reduction (centrifugal and centripetal) do not exclude each other, in line with Harris (2005) and unlike the sharp distinction made by Crosswhite (2004). In the course of vowel reduction and loss in the Finnic languages of Ingria, we observed the elements of both patterns. The mid vowels *o*, *ö* and *e* are raised to the high vowels *u*, *ü* and *i*, which can be seen in the variants *maito/maitu* "milk," *pudro/pudru* "porridge" and *viero/vieru* "wheel" from our questionnaire (Appendix 1), occurring even in the same speaker. At the same time, all vowels could lose their quality completely at later stages and centralise to schwa.

Figure 13 gives a chart of the loss across the vowels of the individual speakers. Both the speakers and the vowels are placed starting with the most conservative and ending with the most innovative ones, which gives the idea of an S-curve of the sound change. The chart shows that the reverse in the ratio between production and categorisation happened in **a* at a much earlier point than in **i* and **u*. Only in the most conservative speaker (IF), the pattern of **a* matches those of the other vowels. The vowel **a* basically jumps over the transitory middle zone with a 50/50 split in production between the old and new values, driven forward by its innovative categorisation. The processes of loss in **i* and **u* run more smoothly. In these two consonant-colouring vowels, in turn, it is mostly an innovative realisation that drives forward the process of change.

4.2 Formation and Loss of Consonantal Palatalisation, Labialisation and Aspiration

Already at stage 1, the vowel qualities revealed significant differences in their patterns of reduction and loss both in production and perception. It seems that vowel loss starts from **i*, and yet is lexically and grammatically stimulated to a large extent (see 3.4). Later on, as loss is generalised throughout the entire phonetic system, **a* takes over as the leader because its reduction and loss proceed faster (it can be seen in speakers starting from CI). This is presumably linked to the strongest qualitative reduction and devoicing of the latter, and also to the fact that **a* does not give any colour (i.e., secondary localisation) to the consonants, unlike **i* and **u*. In **i*, on the other hand, the qualitative reduction is the weakest. For the vowel *u*, no reduction or loss is yet perceived at stage 1, even if qualitative reduction is already very frequent.

An overall trend observed across all vowels and speakers is that during vowel loss the prototypical variants (zero vowel reflexes) of the new category [–SEGMENT] are gaining ground first of all at the cost of the prototypical variants (full clear modal vowels) of the old category [+SEGMENT]. The intermediate variants form a belt which in total accounts for about 20–30% of each sample and slides down the scale. The only major exception to this is a significant cluster of consonantal palatalisation, which replaces a substantial part of the full non-reduced modal vowel *i* between stages 1 and 2. The vowel *i* is the first vowel type to exhibit loss and is nearly as fast in phonetic loss as schwa (<**a*). However, even the stage 2 speakers typically did not yet categorise the reflexes of *i* and *u* as consonantal features. SI1 was consistent in perceiving both as vowels, while SI2 perceived palatalisation also rather as a vowel, but labialisation already rather as loss. This is likely linked to a greater acoustic, articulatory and therefore perceptual salience of *i* and palatalisation compared to *u* and labialisation. Matthies et al. (2001) report the same tendency for quality preservation in *i* even in fast speech. Phonological consonantal palatalisation is apparently formed earlier than labialisation, at least in the history of these varieties, but is lost from the language more slowly. The possible impact of the front vowel harmony characteristic of the Finnic languages of Ingria on this asymmetry is yet to be investigated.

The high level of susceptibility of *u* to qualitative reduction and concomitant loss of rounding might be one of the factors impeding the formation of phonemic consonantal labialisation over loss of the segmental vowel at the later stages of reduction. The vowel *u* is the most conservative of the three vowels in terms of vowel loss, as it retained the largest "vocalic" cluster of realisations in all speakers, but it manifested gradual transitions between the stages in all aspects: qualitative reduction, devoicing and loss. No robust cluster of consonantal labialisation as a trace of *u* is formed; the segmental vowel is rather lost directly. Evidence from vowel perception and neuroimaging (see 1.2) also suggests that *u* is a more complex unit than *i* for brain processing, and acoustically and perceptually less salient.

The difference in size of the consonantal palatalisation and labialisation clusters between stages 2 and 3 could also be attributed to the articulatory properties of these two features. In the Finnic languages of Ingria, consonants do not typically undergo coarticulatory labialisation along their whole length; only the very last portion of the segment is regressively affected. Labialised consonants are often aspirated consonants where just the aspiration portion is labialised rather than the consonant itself (cf. labialised vs. plain aspiration on Fig. 14, 15). The labialised aspiration is then "eaten" away by reduction, and the consonant remains plain.

Consonantal palatalisation (Fig. 16) has a much more powerful impact on the articulation of consonants in these languages. Especially for dentals (and specifically for *l* and *t*), it is a full rather than secondary palatalisation, with a shift of the primary articulation towards the palatal region of the vowel tract (cf. surveys in Kochetov, 2011; Krämer & Urek, 2016). Our preliminary observations show that the number of palatalised consonants and the degree of their

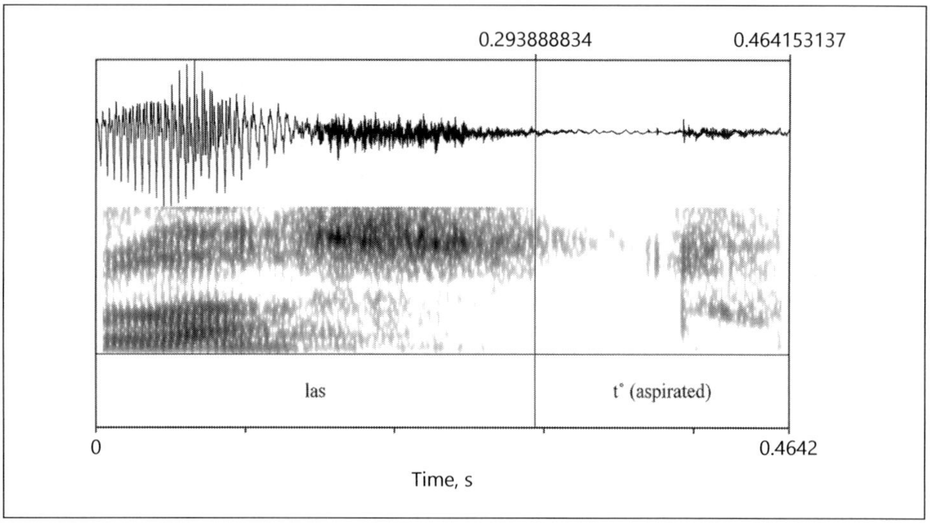

Fig. 14. Aspirated labialised C (**lastu* [łast^{oh}] "chip") PS.

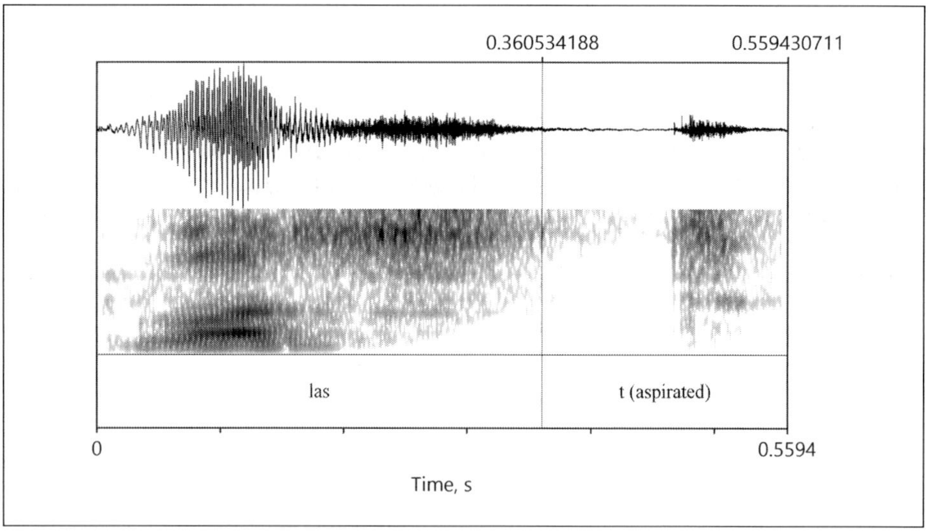

Fig. 15. Aspirated C (**lasta* [łast^{h}] "child:PRT") PS.

palatalisation manifest a positive correlation with the degree of vowel reduction in the Finnic languages of Ingria. The more advanced the vowel reduction and loss, the bigger the number of palatalised consonantal phonemes to be distinguished for any particular variety and the stronger the palatalisation from the phonetic point of view. Ingrian Finnish and Soikkola Ingrian have dental pala-

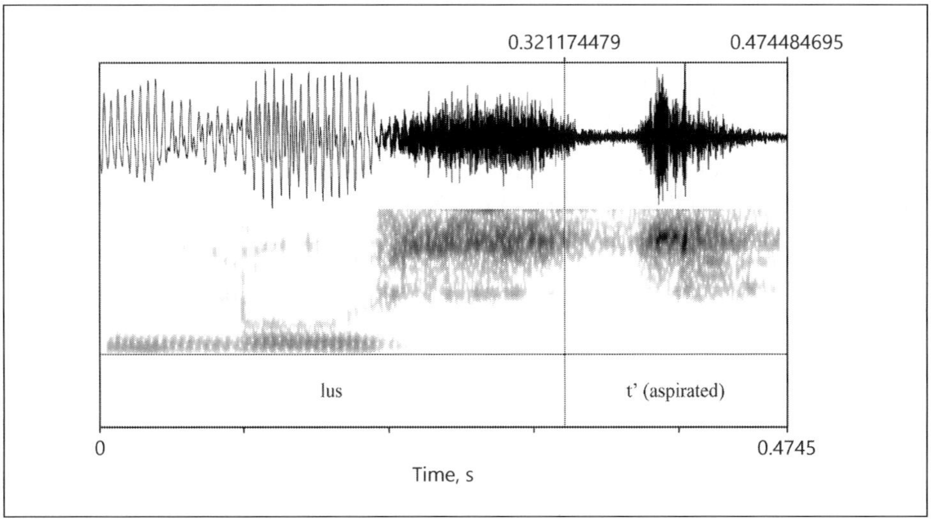

Fig. 16. Aspirated palatalised C (**lusti* [łust^{jh}] "beautiful") AU.

talised phonemes, but in Votic their inventory is significantly larger. At the same time, in Ingrian Finnish, Soikkola Ingrian, Votic and partially Central Lower Luga Ingrian, the consonant *t* is just secondarily palatalised and can be easily realised as a plain one. In most other Lower Luga Ingrian varieties (North, South and partially Central) and in Siberian Ingrian Finnish, in turn, this consonant is always fully palatalised in the context before the high front vowels *i* and *ü*. This palatalisation is so strong that sometimes a palatal stop in pronounced (viz. Leppik, 1975, pp. 116–117; Kuznetsova, 2009, pp. 195–235; Markus & Rožanskij, 2011, pp. 17–18).

These observations are to be further verified, but one could hypothesise that this trend is a result of re-phonologisation. High front vowel quality, which originally stimulated palatalisation, ceases to do so as vowels are progressively reduced and lost. Palatalisation starts to be perceived as a distinct property of consonants and becomes phonetically reinforced. Subsequently, even if the aspiration after consonants disappears, the palatalised articulation in those consonants where it has emerged as a stable property is preserved. Indeed, in the Siberian speaker, consonantal palatalisation was unevenly distributed across consonants. The consonants *l* and *t* were palatalised as a trace of **i* in all cases, and here we can speak about well-formed consonantal palatalisation. The consonants *p*, *k* and *n* manifested palatalisation in 30–60% of the cases, and the consonants *s*, *h*, *r*, *n*, *m* and *v* only sporadically. In these two groups, especially in the last one, a tendency towards complete depalatalisation was observed.

Palatalisation might still, therefore, survive as a phonemic feature, at least for some consonantal types. Labialisation is likely to be lost without any reflexes. Indeed, in neighbouring Estonian, which represents an even more advanced state of the same type of reduction and has passed through the stage of devoiced vowels, consonantal palatalisation as a trace of *i still exists (only for dental consonants and with a trend towards further loss), but no traces of consonantal labialisation were preserved (Teras & Pajusalu, 2014; Kuznetsova, 2016).

A similar disparity in the trajectories of loss of *i and *u is observed in the history of other languages, such as Russian (Šahmatov, 1915, pp. 15–16; Kiparsky, 1963) or Irish (Greene, 1973; Anderson, 2016). Cross-linguistically, labialised consonants are much rarer than palatalised ones. Blevins (2004, p. 204) explains the rarity of certain phonological contrasts through the uncommon occurrence of sound changes giving rise to them. Consonantal palatalisation accounted for 145 cases (3.18%) in PBase (Mielke, 2008; Brohan & Mielke, 2018, pp. 218–219), being one of the most frequent types of sound change, while labialisation included just 38 entries.

We do not know of any cases of emergence of phonemic consonantal aspiration as a result of loss of the following plain schwa, a reflex of *a (nor does PBase give any such examples). This is just an additional indication for the very low perceptual saliency of this schwa, especially in the final position. In our case study, this has led to its fast disappearance both from the mind and the production of the speakers. Nonetheless, one might wonder if at stages 2 and 3 it is still possible to synchronically distinguish a separate phonemic series of aspirated consonants, as opposed to the non-aspirated ones (e.g., *vīn* "bring_away:PRS:1SG" vs. *vīn*^{h} [<**vīna*] "vodka"). At stage 3, this series would be an addition to the plain, palatalised, labialised and labiopalatalised series (Sidorkevič, 2013; Kuznetsova, 2015, 2016). However, all speakers at stages 2 and 3 still perceived reflexes of *i and *u, even if none perceived plain aspiration as a reflex of *a. If the difference between aspirated and plain consonants were still consistently maintained in production, this case would represent an example of a near-merger, the next-to-last step of phonologisation (Barnes, 2006, pp. 223–238; see also 1.1).

5. Conclusions

The novelty of our approach to the problem of correspondence in production and perception at different stages of sound change was that real varieties were studied and speakers assessed words in their own languages. Similar studies have usually involved miniature artificial languages or cross-linguistic assessments, with their own methodological restrictions. The most typical question asked is which of the vowels is perceived, while our enquiry was rather into whether any vowel is perceived at all. If a variety has no literary standard, the latter question is much easier to answer by means of the categorisation test proposed in this study (but only if the intermediary orthography has a segmental principle of encoding).

At the initial stage, vowel reduction and loss are linked to automatisation of execution in production of the old category. Categorisation remains more conservative than production, and the phonetic loss is likely to be concentrated in a few frequent words and grammatical morphemes. At later stages, loss spreads throughout the system and its conditioning becomes purely phonetic. When our speakers pronounced a vowel in more than 70% of the cases, they typically perceived its presence. A decisive qualitative shift seemed to happen in categorisation after the new realisational types had gained more than 70%. The speakers stopped perceiving any segment and categorisation became innovative, while production still showed a certain percentage of vowel preservation and was lagging behind. Reduction is based on a loss of newly meaningless parts, and this last stage may contain near-mergers. Automatisation of execution is likely still at work here, only now the production of the new category is being automatised. Our comparison of the several stages of vowel reduction and loss revealed no irreconcilable contradiction between the two main patterns of reduction, i.e., centrifugal and centripetal. Both were observed in our data: the rise of mid vowels to high vowels and eventual centralisation of all vowels to schwa.

We studied the three basic corner vowels, which are known to share a specific set of properties with regard to reduction and loss, and observed asymmetry in their production and perception. The results suggest two markedness hierarchies among these vowels. As regards the process of reduction and loss itself, the hierarchy of vowels (from the most innovative to the least innovative) is *a* > *i* > *u*. The phonological saliency of the secondary consonantal localisations emerging in the process of loss of these vowels, in turn, would suggest a hierarchy *a* > *u* > *i* (from the least salient to the most salient secondary localisations). Some possible physiological features stimulating both hierarchies were outlined in the paper.

Acknowledgements

I am grateful to Maxim Fedotov for his help with the preliminary segmentation of a part of the data, as well as to Cormac Anderson, Brittany Blankinship, four anonymous reviewers and the associate editors of this Special Issue for their valuable comments.

Disclosure Statement

The authors have no conflicts of interest to declare.

Funding Sources

Data collection and segmentation for this research (2014–2016) were supported by a Grant of the President of the Russian Federation for young PhDs (No. MK-4291.2015.6; "Non-initial vowel reduction in the vanishing Finnic varieties of Ingria").

Author Contributions

Natalia Kuznetsova conceived and performed the study, apart from half of the acoustic analysis of the data and the Praat script for data extraction, which were performed by Vasilisa Verkhodanova.

Appendix 1: List of the Most Typical Carrier Words

T	A	I	U/O	R	A	I	U/O
t	*musta* "black"	*lusti* "beautiful"	*lastu* "chip"	l	*suola* "salt"	*stuoli* "chair"	*škoulu* "school"
t	*vihta* "bath broom"	*risti* "(a) cross"	*lintu* "bird"	l	*naula/nāgla* "nail"	*hīli* "coal"	*laulu* "song"
t	*lūta* "broom"	*puoti* "shop"	*maito/maitu* "milk"	l	*muila* "soap"	*kieli* "tongue"	*joulu* "Christmas"
p	*hāpa* "aspen"	*sīpi* "wing"	*rūpo* "rubbish"; (*ei*) *korpu* "(does not) dry out:3SG"	r	*koira* "dog"	*sūri* "big"	*pudru/pudro* "porridge"
p	*kumpa* "which of the two"	*krāpi* "comb (wool):IMP"	*urpo* "willow"	r	*nuora* "rope"	*nuori* "young"	*vieru/viero* "wheel"
k	*jalka* "leg, foot"	*panki* "bucket"	*hanko* "snowbank"	r	*hāra* "branch"	*hīri* "mouse"	*viiru* "stripe"
k	*nahka* "skin"	*poski* "cheek"	*pehko* "bush"	n	*vīna* "vodka"	*pieni* "small"	*hieno* "fine"
k	*poika* "boy"	*hauki* "pike"	*riuku* "pole"	n	*sauna* "sauna"	*sāni* "sleigh"; *sieni* "mushroom"	*kehno* "worn out"
k	*velka* "debt"	*olki* "straw"	*halko* "billet"	m	*ilma* "air"	*tormi* "storm"	*silmu* "lamprey"
s	*oksa* "twig"	*lapsi* "child"	*paksu* "thick"	m	*surma* "death"	*sormi* "finger"	*formu* "(a) form"
s	*klāsa* "class"; *vatsa* "stomach"	*ūsi* "new"	*haisu* "(a) smell"	m	*māma* "mother"	*Suomi* "Finland"	*solmu* "knot"
h	*vaiha* "change:IMP"	*rīhi* "drying barn"; *jouhi* "horsehair"	*jauho* "wheat"	v	*hīva* "yeast"	*talvi* "winter"	*kaivo* "(a) well"
h	*laiha* "lean, thin"; *turha* "vain"	*tuohi* "birchbark"	*karhu* "(a) bear"; *kaiho* "damage"	v	*līva* "sand"	*hirvi* "elk"; *sarvi* "horn"	*koivu* "birch"

Morphophonological front-vowel stems are underlined.

References

Anderson, C. (2016). *Consonant colour and vocalism in the history of Irish* (Unpublished PhD dissertation), Adam Mickiewicz University in Poznań, Poland.

Baker, A., Archangeli, D., & Mielke, J. (2011). Variability in American English s-retraction suggests a solution to the actuation problem. *Language Variation and Change, 23*(3), 347–374. https://doi.org/10.1017/S0954394511000135

Barnes, J. (2006). *Strength and weakness at the interface: Positional neutralisation in phonetics and phonology*. Berlin, Germany; New York, NY: Mouton de Gruyter.

Baudouin de Courtenay, J. N. (1895). *Versuch einer Theorie phonetischer Alternationen: Ein Capitel aus der Psychophonetik*. Strassburg: K. J. Trübner.

Blevins, J. (2004). *Evolutionary phonology: The emergence of sound patterns*. Cambridge: Cambridge University Press. https://doi.org/10.1017/CBO9780511486357

Blythe, R. A., & Croft, W. (2012). S-curves and the mechanisms of propagation in language change. *Language, 88*(2), 269–304. https://doi.org/10.1353/lan.2012.0027

Brohan, A., & Mielke, J. (2018). Frequent segmental alternations in P-base 3. In L. Hyman & F. Plank (Eds.), *Phonological typology* (pp. 196–228). Berlin, Germany; Boston, MA: Walter de Gruyter. https://doi.org/10.1515/9783110451931-006

Bybee, J. (2001). *Phonology and language use*. Cambridge: Cambridge University Press. https://doi.org/10.1017/CBO9780511612886
Bybee, J. (2015). *Language change*. Cambridge: Cambridge University Press. https://doi.org/10.1017/CBO9781139096768
Bybee, J., File-Muriel, R. J., & de Souza, R. N. (2016). Special reduction: A usage-based approach. *Language and Cognition, 8*(3), 421–446. https://doi.org/10.1017/langcog.2016.19
Crosswhite, K. (2004). Vowel reduction. In B. Hayes, R. Kirchner, & D. Steriade (Eds.), *Phonetically-based phonology* (pp. 191–231). Cambridge: Cambridge University Press. https://doi.org/10.1017/CBO9780511486401.007
Davis, T., & Poldrack, R. A. (2014). Quantifying the internal structure of categories using a neural typicality measure. *Cerebral Cortex (New York, N.Y.: 1991), 24*(7), 1720–1737. https://doi.org/10.1093/cercor/bht014
de Lacy, P. (2006). *Markedness: Reduction and preservation in phonology*. Cambridge: Cambridge University Press. https://doi.org/10.1017/CBO9780511486388
Delattre, P. (1969). An acoustic and articulatory study of vowel reduction in four languages. *International Review of Applied Linguistics in Language Teaching, 7*(4), 295–325. https://doi.org/10.1515/iral.1969.7.4.295
Eulitz, C., & Obleser, J. (2007). Perception of acoustically complex phonological features in vowels is reflected in the induced brain-magnetic activity. *Behavioral and Brain Functions, 3*(1), 26. https://doi.org/10.1186/1744-9081-3-26
Flemming, E. (1995). *Auditory representations in phonology* (Unpublished PhD dissertation), UCLA, USA.
Flemming, E. (2004). Contrast and perceptual distinctiveness. In B. Hayes, R. Kirchner, & D. Steriade (Eds.), *Phonetically-based phonology* (pp. 232–276). Cambridge: Cambridge University Press. https://doi.org/10.1017/CBO9780511486401.008
Greene, D. (1973). The growth of palatalisation in Irish. *Transactions of the Philological Society, 72*(1), 127–136. https://doi.org/10.1111/j.1467-968X.1973.tb01017.x
Gureckis, T. M., & Goldstone, R. L. (2008). The effect of the internal structure of categories on perception. In L. Shi, N. H. Feldman, & T. L. Griffiths (Eds.), *Proceedings of the 30th annual conference of the Cognitive Science Society* (pp. 1876–1881). Austin, TX: Cognitive Science Society.
Hall, K. C., Hume, E., Jaeger, F. T., & Wedel, A. (2018). The role of predictability in shaping phonological patterns. *Linguistics Vanguard*, 20170027.
Hanique, I., Ernestus, M., & Boves, L. (2015). Choice and pronunciation of words: Individual differences within a homogeneous group of speakers. *Corpus Linguistics and Linguistic Theory, 11*(1), 161–185. https://doi.org/10.1515/cllt-2014-0025
Harris, J. (2005). Vowel reduction as information loss. In P. Carr, J. Durand, & C. J. Ewen (Eds.), *Headhood, elements, specification and contrastivity: Phonological papers in honour of John Anderson* (pp. 119–132). Amsterdam, The Netherlands: John Benjamins Publishing. https://doi.org/10.1075/cilt.259.10har
Hay, J., & Foulkes, P. (2016). The evolution of medial /t/ over real and remembered time. *Language, 92*(2), 298–330. https://doi.org/10.1353/lan.2016.0036
Hyman, L. M. (1976). Phonologisation. In A. Juilland (Ed.), *Linguistic studies presented to Joseph H. Greenberg* (pp. 407–418). Saratoga: Anma Libri.
Johnson, K. (2015). Vowel perception asymmetry in auditory and phonemic listening. *UC Berkeley Phonology Lab Annual Report 2015*. Available from https://escholarship.org/uc/item/21t337g
Kapatsinski, V. (2018). *Changing minds changing tools: From learning theory to language acquisition to language change*. Cambridge, MA: MIT Press.
Kiparsky, V. (1963). *Russische historische Grammatik. Band I: Die Entwicklung des Lautsystems*. Heidelberg, Germany: Universitätsverlag Winter Heidelberg.
Kirby, J. P. (2010). *Cue selection and category restructuring in sound change* (Unpublished PhD dissertation), University of Chicago, USA.
Kirchner, R. M. (1998). *An effort-based approach to consonant lenition* (Unpublished PhD dissertation), UCLA, USA.
Klatt, D. H., & Klatt, L. C. (1990). Analysis, synthesis, and perception of voice quality variations among female and male talkers. *The Journal of the Acoustical Society of America, 87*(2), 820–857. https://doi.org/10.1121/1.398894
Kochetov, A. (2011). Palatalisation. In M. van Oostendorp, C. J. Ewen, K. Rice, & E. V. Hume (Eds.), *The Blackwell companion to phonology* (Vol. 3, pp. 1666–1690). London, UK: John Wiley & Sons.

Krämer, M., & Urek, O. (2016). Perspectives on palatalisation. *Glossa: A Journal of General Linguistics, 1*(1):31. http://doi.org/10.5334/gjgl.226

Kruszewski, N. V. (1883). *Očerk nauki o jazyke.* Kazan, Russia.

Kuznetsova, N. (2009). *Fonologičeskie sistemy ižorskih dialektov* (Unpublished PhD dissertation), Saint Petersburg: Institut lingvističeskih issledovanij RAN, Russia.

Kuznetsova, N. (2011). Typology of prosodic systems in Low Luga Izhorian varieties. In P. K. Austin, O. Bond, M. Charette, D. Nathan, & P. Sells (Eds), *Proceedings of Conference on Language Documentation and Linguistic Theory 2* (pp. 185–192). London, UK: SOAS.

Kuznetsova, N. (2012a). Prosodika slovoformy v nižnelužskom dialekte ižorskogo jazyka. *Acta linguistica Petropolitana, VIII*(1):43–103.

Kuznetsova, N. (2012b). Vowel reduction in Lower Luga Ingrian: Scientific description and "folk" perception. In G. V. Golovko, M. L. Lurie, M. Z. Muslimov, & S. A. Oskolskaya (Eds.), *"Folk linguistics": Language from speakers' perspective. Paper abstracts for the conference, St. Petersburg, November 19–21, 2012* (pp. 85–87). St. Petersburg, Russia: Nestor-Istorija.

Kuznetsova, N., & Sidorkevič, D. (2012). Priloženie 6: Karty k statjam. *Acta linguistica Petropolitana, VIII*(1):559–567.

Kuznetsova, N. (2015). Two phonological rarities in Ingrian dialects. In M. Hilpert, J. Duke, C. Mertzlufft, J.-O. Östman, & M. Rießler (Eds.), *New trends in Nordic and general linguistics* (pp. 91–117). Berlin, Germany: De Gruyter.

Kuznetsova, N. (2016). Evolution of the non-initial vocalic length contrast across Finnic varieties of Ingria and the adjacent areas. *Linguistica Uralica, 52*(1), 1–25. https://doi.org/10.3176/lu.2016.1.01

Kuznetsova, N., Markus, E., & Muslimov, M. (2015). Finnic minorities of Ingria: The current sociolinguistic situation and its background. In H. Marten, M. Riessler, J. Saarikivi, & R. Toivanen (Eds.), *Cultural and linguistic minorities in the Russian Federation and the European Union* (pp. 127–167). Berlin: Springer.

Laanest, A. (1980). Inkeroismurteiston suhteista suomen kaakkoismurteisiin. *Virittäjä, 84*, 142–149.

Labov, W. (1999). *Principles of linguistic change. Vol. 1: Internal factors.* London, UK: Blackwell.

Labov, W., Karen, M., & Miller, C. (1991). Near-mergers and the suspension of phonemic contrast. *Language Variation and Change, 3*(1), 33–74. https://doi.org/10.1017/S0954394500000442

Laver, J. (1994). *Principles of phonetics.* Cambridge: Cambridge University Press. https://doi.org/10.1017/CBO9781139166621

Leppik, M. (1975). *Ingerisoome kurgola murde fonoloogilise süsteemi kujunemine.* Tallinn, Estland: Eesti NSV Teaduste Akadeemia, Keele ja Kirjanduse Instituut.

Leskinen, H. (1973). Kaakkoissuomalaisen loppuheiton yleisyydestä ja alkuperästä. *Suomalais-Ugrilaisen Seuran Aikakauskirja, 72*, 210–221.

Lindblom, B. (1963). Spectrographic study of vowel reduction. *The Journal of the Acoustical Society of America, 35*(11), 1773–1781. https://doi.org/10.1121/1.1918816

Lindblom, B. (1990). Explaining phonetic variation: A sketch of the H&H theory. In W. J. Hardcastle & A. Marchal (Eds.), *Speech production and speech modelling* (pp. 403–439). Dordrecht, The Netherlands: Springer. https://doi.org/10.1007/978-94-009-2037-8_16

Mägiste, J. (1925). *Rosona (Eesti Ingeri) murde pääjooned.* Tartu, Estonia.

Manca, A. D., & Grimaldi, M. (2016). Vowels and consonants in the brain: Evidence from magnetoencephalographic studies on the N1m in normal-hearing listeners. *Frontiers in Psychology, 7*, 1413. https://doi.org/10.3389/fpsyg.2016.01413

Markus, E., & Rožanskij, F. (2011). *Sovremennyj vodskij jazyk. Teksty i grammatičeskij očerk. Vol. 2.* Saint Petersburg, Russia: Nestor-Istorija.

Matthies, M., Perrier, P., Perkell, J. S., & Zandipour, M. (2001). Variation in anticipatory coarticulation with changes in clarity and rate. *Journal of Speech, Language, and Hearing Research: JSLHR, 44*(2), 340–353. https://doi.org/10.1044/1092-4388(2001/028)

Maye, J., & Gerken, L. (2000). Learning phonemes without minimal pairs. In S. C. Howell, et al. (Eds.), *Proceedings of the Annual Boston University Conference on Language Development* (Vol. 24, pp. 522–533). Somerville, MA: Cascadilla Press.

Mielke, J. (2008). *The emergence of distinctive features.* Oxford: Oxford University Press. PBase available from http://pbase.phon.chass.ncsu.edu [last accessed 01.10.2018].

Mompeán-González, J. A. (2004). Category overlap and neutralisation: The importance of speakers' classifications in phonology. *Cognitive Linguistics, 15*(4), 429–469. https://doi.org/10.1515/cogl.2004.15.4.429

Muslimov, M. (2005). *Jazykovyje kontakty v Zapadnoj Ingermanlandii (nižnee tečenije reki Lugi)* (Unpublished PhD dissertation), Sankt-Peterburg: Institut lingvističeskih issledovanij RAN, Russia.
Nirvi, R. E. (1972). Siperian inkeriläisten murteesta ja alkuperästä. *Kotiseutu, 2/3*, 92–95.
Ohala, J. J. (1989). Sound change is drawn from a pool of synchronic variation. In L. E. Breivik & E. H. Jahr (Eds.), *Language change: Contributions to the study of its causes* (pp. 173–198). Berlin, Germany: Walter de Gruyter. https://doi.org/10.1515/9783110853063.173
Olejarczuk, P., Kapatsinski, V., & Baayen, R. H. (2018). Distributional learning is error-driven: The role of surprise in the acquisition of phonetic categories. *Linguistics Vanguard: Multimodal Online Journal, 4*(s2, S2). https://doi.org/10.1515/lingvan-2017-0020
Pearce, M. (2008). Vowel harmony domains and vowel undershoot. *UCL Working Papers in Linguistics, 20*:115–140.
Polka, L., & Bohn, O. S. (2003). Asymmetries in vowel perception. *Speech Communication, 41*(1), 221–231. https://doi.org/10.1016/S0167-6393(02)00105-X
Polka, L., & Bohn, O. S. (2011). Natural Referent Vowel (NRV) framework: An emerging view of early phonetic development. *Journal of Phonetics, 39*(4), 467–478. https://doi.org/10.1016/j.wocn.2010.08.007
Priva, U. C. (2017). Informativity and the actuation of lenition. *Language, 93*(3), 569–597. https://doi.org/10.1353/lan.2017.0037
Ritter, S., Mücke, D., & Grice, M. (2018, October 2). *The dynamics of intonation: Categorical and continuous variation in an attractor-based model.* https://doi.org/10.31234/osf.io/qkm6b
Roettger, T. B., Winter, B., Grawunder, S., Kirby, J., & Grice, M. (2014). Assessing incomplete neutralisation of final devoicing in German. *Journal of Phonetics, 43*, 11–25. https://doi.org/10.1016/j.wocn.2014.01.002
Rosch, E. (1978). Principles of categorisation. In E. Rosch & B. B. Lloyd (Eds.), *Cognition and categorisation* (pp. 27–48). Hillsdale, NJ: Lawrence Erlbaum Associates.
Šahmatov, A. A. (1915). *Očerk drevneišego perioda istorii russkogo jazyka.* Petrograd, Russia.
Sidorkevič, D. V. (2013). *Jazyk ingermanlandskih pereselencev XIX v. v Sibiri (struktura, dialektnye osobennosti, kontaktnye javlenija)* (Unpublished PhD dissertation), Sankt-Peterburg: Institut lingvističeskih issledovanij RAN, Russia.
Stevens, K. N. (2004). Invariance and variability in speech: Interpreting acoustic evidence. In J. Slifka, S. Manuel, & M. Matthies (Eds), *From sound to sense: 50+ years of discoveries in speech communication* (pp. B-77–B-85). Cambridge, MA: Research Laboratory of Electronics, MIT.
Stevens, M., & Harrington, J. (2014). The individual and the actuation of sound change. *Loquens, 1*(1), e003. https://doi.org/10.3989/loquens.2014.003
Szeredi, D. (2010). Vowel centralisation and vowel harmony in Hungarian. *The Odd Yearbook, 8*, 111–137.
Teras, P., & Pajusalu, K. (2014). Palatalisatsioonist ja prepalatalisatsioonist spontaanses eesti keeles. *Keel ja Kirjandus, 4*, 257–269.
Traunmüller, H., & Öhrström, N. (2007). Audiovisual perception of openness and lip rounding in front vowels. *Journal of Phonetics, 35*(2), 244–258. https://doi.org/10.1016/j.wocn.2006.03.002
Uchanski, R. M. (2005). Clear speech. In D. B. Pisoni & R. E. Remez (Eds.), *The handbook of speech perception* (pp. 207–235). Oxford: Blackwell. https://doi.org/10.1002/9780470757024.ch9
Vallabha, G. K., McClelland, J. L., Pons, F., Werker, J. F., & Amano, S. (2007). Unsupervised learning of vowel categories from infant-directed speech. *Proceedings of the National Academy of Sciences of the United States of America, 104*(33), 13273–13278. https://doi.org/10.1073/pnas.0705369104
van Bergem, D. R. (1995). Perceptual and acoustic aspects of lexical vowel reduction, a sound change in progress. *Speech Communication, 16*(4), 329–358. https://doi.org/10.1016/0167-6393(95)00003-7
van der Hulst, H. (2018). *Asymmetries in vowel harmony: A representational account.* Oxford: Oxford University Press. https://doi.org/10.1093/oso/9780198813576.001.0001
Vatakis, A., Maragos, P., Rodomagoulakis, I., & Spence, C. (2012). Assessing the effect of physical differences in the articulation of consonants and vowels on audiovisual temporal perception. *Frontiers in Integrative Neuroscience, 6*, 71. https://doi.org/10.3389/fnint.2012.00071
Walker, R. (2011). *Vowel patterns in language.* Cambridge: Cambridge University Press. https://doi.org/10.1017/CBO9780511973710
Wanrooij, K., Escudero, P., & Raijmakers, M. E. (2013). What do listeners learn from exposure to a vowel distribution? An analysis of listening strategies in distributional learning. *Journal of Phonetics, 41*(5), 307–319. https://doi.org/10.1016/j.wocn.2013.03.005
Yu, A. C. (2013). Socio-cognitive processing and the actuation of sound change. In A. C. Yu (Ed), *Origins of sound change: Approaches to phonologisation* (pp. 201–227). Oxford: Oxford University Press.

Phonetica

Author Index Vol. 76, No. 2–3, 2019

KARGER

E-Mail karger@karger.com
www.karger.com/pho